D0709473

Farm Families & Change

Farm Families & Change

in Twentieth-Century America

MARK FRIEDBERGER

THE UNIVERSITY PRESS OF KENTUCKY

Scholarly publisher for the Commonwealth,
serving Bellarmine College, Berea College, Centre
College of Kentucky, Eastern Kentucky University,
The Filson Club, Georgetown College, Kentucky
Historical Society, Kentucky State University,
Morehead State University, Murray State University,
Northern Kentucky University, Transylvania University,
University of Kentucky, University of Louisville,
and Western Kentucky University.

Editorial and Sales Offices: Lexington, Kentucky 40506-0024

Library of Congress Cataloging-in-Publication Data

Friedberger, Mark.
Farm families and change in twentieth-century
America.

Bibliography: p.
Includes index.
1. Family farms—United States. 2. Family—United
States. 3. United States—Rural conditions. 4. Agri-
culture—Economic aspects—United States. I. Title.
HD1476.U5F75 1988 338.1'6 87-34548
ISBN 0-8131-1636-8

This book is printed on acid-free paper meeting
the requirements of the American National Standard for
Permanence of Paper for Printed Library Materials. ♾

To the Farm Families
of Iowa and the Central Valley

Contents

Tables

Acknowledgments

This book would not have been completed without the support of two remarkable federal agencies, the National Institutes of Health and the National Endowment for the Humanities. I would like to thank the peer reviewers, who saw some merit in my proposals to study farm families, and the staffs at the National Institute of Child Health, the National Institute on Aging, and the Research Division of the National Endowment, whose aid allowed me to get on with the work in total freedom.

At my home base in Chicago, the staff of the library at the University of Illinois at Chicago were supportive, as was the Department of History, and the Office of Sponsored Research. Richard Jensen read a first draft of the manuscript, made some perceptive comments, and as always was encouraging. Tim and Kathy Lennon sustained me at a difficult time and were most generous with their hospitality.

I would like to thank the many Iowans who helped smooth the way over a long period. Tex Heyer, Jim Graham, Helen Frieden, and her staff at the Recorder's Office in West Union were always welcoming. Deb and Don Bahe, Sandy and Gordon Murry-John, and Larry Harris and Denise O'Brien tolerated my sudden appearance on a number of occasions and entertained me royally. The Reverend David Ostendorf was generous with his time and allowed me access to the activities of his organization. Lorraine Kuennen introduced me to her community, and Nick Kuennen opened his home to me. Down the years the staff of the Benton County Recorder, and Francis Beggs and Chuck Juhl at Benton County Title were generous with their assistance.

In California, too, a number of individuals and institutions went out of their way to be helpful. The staff of Special Collections, California State University at Fresno; Fresno County Public Library; Kings County Public Library; the Clerk and Recorder of Kings County; the staff of Kings County Title; the staff of Safeco Title in Visalia; and the library of the Fresno *Bee*, all were most accommodating. Barbara Saville, Stuart Bartlett, Joe Cotta, Jim Reynolds, and Jonathan and Nancy

Barker contributed in countless ways to make my time in San Joaquin Valley as productive as possible.

Portions of chapter 4 appeared in an earlier form in "The Farm Family and the Inheritance Process: Evidence from the Cornbelt, 1870-1950," *Agricultural History* 57 (1983): 1-13; and "Handing Down the Home Place: Farm Inheritance Strategies in Iowa 1870-1945," *Annals of Iowa* 47 (1984): 518-36. I thank the publishers for permission to use that material here.

I would like to thank my editors, especially Judith Bailey, for skillful editorial work. My children, Matthew and Eleanor, stoically endured my long absences over the period it took to complete this book and never complained. Finally, Friar Tuck, who logged thousands of miles around Iowa, behaved in exemplary fashion during many visits to farm families. It is to the latter that this book is dedicated.

Introduction

In recent years Americans have been bombarded with depictions of hard times for farm families. Movies, network news coverage, background features in magazines, and even extravaganzas like the "Farm Aid" concerts have focused on the human dislocation, emotional suffering, and adverse economic fallout on the farm. Hollywood and media accounts of bankruptcies and foreclosures grimly portrayed the struggle of farm families to remain on the land. Thanks to the intimacy of television and the cinema, the public glimpsed a phenomenon that seemed like news. Actually, however, farmers are facing a situation common in the American past: adjustment, in circumstances beyond their control, of a rural population to the realities of a changing economy. The migration of country people to other occupations and localities and the surrender to a new way of life that substitutes for the rhythms of nature the timetables of shop and factory are major themes of social history. The situation of the modern farm family, forced to sacrifice the farm which is home, livelihood and legacy to their children, is symptomatic of a continuous adjustment of farming since the time the land was first settled.

In the winter of 1921-1922, for instance, the farm economy was still in a deep recession after the sharp downturn of 1920. The secretary of agriculture, Henry C. Wallace quoted as typical a pathetic scene brought to his attention by a friend: "Our neighbor joining us on the east, a hardworking man, had rented 320 acres of land. He and his wife and one hired man farmed it. They had about 100 head of cattle and about the same number of hogs. The 1st of December they turned everything over to the landlord, save one team, which they hitched to an old wagon, put in their household goods, got on the wagon themselves, and drove away to town to get work at day labor and make a new start in life."[1]

In this book I intend to apply a historical perspective to the problem of the survival of the family farm from the turn of the century to the present. Concern for the "waning of American Gothic" is natural and understandable.[2] The family farm is an icon of American life, a unique organization that combines workplace and dwelling. The family that lives together and works as a unit has a number of advantages, including intergenerational access to credit and advice, the diffusion of risk, and a captive labor supply. On the other hand, many family farms have foundered because of intrafamily squabbles over management decisions, poor utilization of family labor, and a reluctance on the part of the older generation to allow children to become educated and to adopt new technology.

The problems that plagued family farmers in the twenties and in the eighties, symbolized by bankruptcies and foreclosures, are symptoms of an economic crisis as it hit individual farms. Often, the emotional, and human side of the failure of a family farm is emphasized, but these failures reflect the complex interrelationships of banks, corporations, national and international economic transformations, federal and local politics, and community institutions and businesses. Thus, any investigation into the condition of the family farm needs to explore the forces that have impinged on it during the twentieth century, particularly those relevant to the uncertain period of the eighties.

At a time when farm policy and, by implication, farming methods are the subject of scrutiny, it is worthwhile to survey the impact of structural change on the actual process of growing crops in the past. And there are a number of other themes in the history of American agriculture that have important implications for the farm crisis—for example, who controlled the land and had access to it and how farm families maintained continuity in their operations over time. Given the critical role credit has played in the downfall of many farmers in the past few years, a thorough historical analysis of farm credit mechanisms is long overdue. Finally, the farm crisis has shown the vulnerability not only of the farm family as a unit of production but also of the local service community that depends on the custom of the farm family for its economic health. I have, therefore, approached the study of the family farm in the eighties through a historical understanding of such themes as farming methods, tenure, inheritance, credit, and family and community dynamics.

In this book I want to show how the farm family has been able to use available resources to ensure survival. Successful families have looked at farming as an occupation requiring a great deal of commitment. They have sought to own land and then to enlarge their holdings, and they have handed the farm down to the next generation. They have generally been conservative in their use of credit, and have preferred a diversified farm to a specialized one. Even with the vast structural changes in farming in the past few decades, such qualities have remained important and have, in some measure, enabled certain families and communites to remain strong during the downturn of the eighties.

Ideally the entire country should be studied, but logistics and cost considerations have permitted only a case study approach. The setting is the classic farm country of the Iowa corn belt and the southern Central Valley of California. As two of the most important farm areas of the country where farm families are still engaged in production agriculture, they provide reasonable models for the state of farming in the rest of the nation.[3] Farmers in these two regions share a common heritage in the way they go about their day-to-day operations.

To be sure, there are obvious contrasts between Iowa and the southern San Joaquin. The heavy dependence on farm labor, the scale of operations, and above all the hydraulic society of the Central Valley make the corn belt look pristine in comparison.[4] In Iowa the image of red barns and white farm houses dotted along unpaved roads conjures up a bucolic past and supposedly generates a sentimental attachment of the midwestern farm family to their land, whereas the typical Central Valley dairy farm consists of a functional split-level bungalow surrounded by corrals and shelters. But the differences go much deeper. The legacy of industrialized agriculture in the Central Valley has necessarily left a mark on its society, giving the California rancher a detached, businesslike, almost urban orientation not to be found in the corn belt. The impact of industrialized agriculture in California has been felt in irrigation policies, which have never conformed to nineteenth century stipulations that federal subsidies be evenly distributed and that water policy benefit the small farmer. For years industrialized agriculture has been dependent on legally and illegally imported labor, which has tended to weaken and impoverish communities and to divide them along class and ethnic lines. And finally industrialized agriculture has degraded the environment, damaging soil, water, and air quality.[5] In Iowa, by contrast, while modern agricultural practices have caused severe erosion and other environmental problems, the other legacies of industrialized agriculture have made less impact.

Inevitably, then, the character of farming in the two areas differs. Yet, though due notice will be taken of these differences, it is important to recognize that at the level of the farm family unit, management decisions, questions involving inheritance, and reorganization—the very stuff of the farm business—have similar ramifications whether they are made in Kings County, California, or Benton County, Iowa. Even in other areas, California is not altogether exceptional.[6] The influence of the Midwest is strong in many communities in the Central Valley. To a certain extent some resemble midwestern farm towns where, but for the winter fogs, the sun shines twelve months in the year. Nor did culture diffuse itself entirely from the Midwest to the

West. The term *agribusiness* was coined by a Fresno farm chemical executive in the 1940s.[7] The modernization of agriculture in the corn belt, especially in the all-important service sector, has borrowed heavily on ideas and organizations first perfected in the Central Valley.

The nature of farming in California makes the field of study problematic. Social scientists usually say that in a family farm the control of the decision-making process rests with the family; the farm provides the primary source of income, and most of the labor and working capital (not necessarily the land) is supplied by family members.[8] Such a definition makes sense in the midwestern corn belt, but in the Central Valley of California quite small operations have always required substantial amounts of outside labor to supplement that of family members. For this reason, I have stretched the definition of family farms to include those managed by family members, but having nonrelatives in subordinate positions.

The Impact of Boom and Bust

More than urban America the farm economy has been susceptible to severe cycles of boom and bust. These have had a major impact on farm operations, tenure, credit, and the health and wealth of the farm family itself and the community in which it lived. The farm crisis of the eighties, then, takes its place in a long line of farm depressions of the 1890s, the 1920s, and the 1930s, when farmers as price takers not price makers were confronted with returns so low that they could not meet their credit obligations. Recurring overproduction aggravated this cycle of farm recessions, and farmers reacted to them with a series of agrarian protests, embracing voluntary associations, politics, and on some occasions direct action to try to solve their economic problems.

The cooperative activity of the Farmer's Alliance from the 1880s to the 1890s and the political successes of the Populist party, which grew out of the grass-roots organizing of the alliance, were tributes to the ability of farmers to make an impression on the political economy of America in the late nineteenth century. Equally characteristic however, was the inability of farmers to persevere in their efforts; the promising beginnings soon withered. Farms were prosperous in the first two decades of this century, but again in 1920 a sharp recession closed out a speculative land boom, with major impact on the farm economy for the rest of the decade in corn-belt states like Iowa. By this time, however, political support of farm causes had achieved respectability. In North Dakota and Minnesota farmer-labor alliances controlled state politics. On the national scene the "farm bloc," composed of resurgent midwestern legislators, tried to force through Congress a series of measures aimed at solving the problems of low prices and excess production. Unfortunately, these proposals met with resistance from both the Coolidge and the Hoover administrations.[10]

The advent of the Great Depression of the 1930s ensured that agricultural problems received primary attention from President Roosevelt. In 1933, while the new administration was struggling to put together its farm program, farmers in some areas of the Iowa corn belt initiated their own attempts to withhold farm products. In northwest Iowa, where the Farmers' Holiday Association was strongest, the violence was serious enough to prompt the mobilization of the National Guard.[11] Across the continent in Kings and Tulare counties, California, in October 1933, a cotton strike erupted. Farm laborers held out for a number of weeks, and in the end won the increased pay they had sought. This action was made possible by New Deal labor legislation that permitted collective bargaining by workers.[12] Although the growers lost the first round with labor, however, they rapidly mobilized their forces, and it was not until the sixties that labor again challenged the status quo.

The Great Depression was a watershed for agriculture. During these years, Congress passed a vast amount of legislation to benefit farming, culminating, after a series of false starts, in the Agricultural Adjustment Act of 1938. This measure, which provided a base for all farm programs for the next half century, included soil conservation measures, acreage allotments, storage provisions, marketing, crop quotas, and insurance and was designed to support the price of farm commodities by controlling production and managing supplies.[13]

Over the following four decades the farm programs acted as a safety net for agriculture to prevent the kind of collapse that had occured in the thirties. Gradually the programs became institutionalized. Such was the political economy of agriculture that by the eighties they virtually operated as entitlement programs, under which the government was obligated to pay whatever it cost to assure farmers the benefits for which they were qualified. Agribusiness, particularly the grain trade, was especially well placed to take advantage of the programs. A close relationship grew between government and agribusiness, manifested in the ease with which secretaries of agriculture from either political party moved from the Department of Agriculture to high-paying jobs in agribusiness and food processing firms after their period of government service was over.

The inclusion in the farm programs of only certain commodities—wheat, feed grains, dairy, cotton, tobacco, sugar, rice, and peanuts, were the major ones—had particular significance for California agriculture, which for years maintained a special fondness for the free market. By the seventies, California ranchers could grow over two hundred different commodities, most of them not included in the program. Californians, therefore, resented the way agricultural policy was dictated by powerful politicians from the Midwest and particularly the South, who were able to advance the interests of their farm constituants by virtue of their senior committee positions in Congress. One California farm editor summarized this attitude succinctly. "The Farm Bills of the past two decades," he wrote, "have been miserable affairs written to protect substandard farms, and in some cases turning into relief programs that helped price us out of the world market."[14] Ironically, just as this was being written, because of the economic downturn, California farmers were scrambling to enroll under the 1985 farm bill.

The number of people engaged in agriculture reached its high point in the thirties; a long slow exodus from farming began. Small operators were pushed from the land, and their neighbors bought their farms to enlarge their operations. Other families and farm laborers willingly left farming for easier jobs and better pay in towns and cities. Most who left in the Midwest stayed within easy reach of the old farm, partly because they liked rural living and also to stay close to relatives and friends.[15] The memory of the thirties produced a collective mentality in farm families that stressed frugality. It dominated the thinking of a whole generation, producing resistance to innovation, but gradually in the forties and fifties modernization altered attitudes. Informal government policy encouraged "family and community systems to adopt more specialized, impersonal, and outgoing" attitudes.[16] Greater attention was paid to making farms "efficient," by substituting capital, in the form of machinery, for labor. A system based on mechanized efficiency viewed labor as more expensive than capital. For this reason, the reduction of labor was seen as inherently efficient and allowed farmers to expand while lowering consumer costs. Such ideas probably reached their apogee in the middle seventies when the Department of Agriculture under Earl Butz endorsed the notion that the best of all possible farms produced the most goods with the least labor.[17]

Origins of the Farm Crisis

The seventies were an aberration from the usual course of development in farming. The world food crisis of 1972-1973 made agriculture strategically important, and most farmers, not surprisingly, felt comfortable with a philosophy that supported the exploitation of overseas markets and the maximization of production to meet the demands of foreign trade. The bonanza of fencerow-to-fencerow planting silenced any alternative philosophy within the farming community. Growth, after all, was what business was about. If nothing else, the purchase of more land and larger machinery and the upgrading of buildings and homes brought a heightened sense of purpose to farming. Higher net worth worked wonders for the morale of families who in the seventies were able to live for the first time at a standard approximating that of the business people in town.[18]

When farming thus became a growth industry, the agricultural exodus was partly arrested. Nationally during the decade around seventy thousand people under thirty-five either inherited or bought land and entered farming. The central feature of these years was the enormous increase in the price of agriculture's major asset, its land; farmers began to be preoccupied with capital gains rather than income. Nationally, between 1970 and 1981, the price of an acre of farmland rose from three hundred to seventeen hundred dollars. In some areas of the corn belt it was not uncommon for farms to sell for four thousand dollars an acre, and in the Central Valley fruit orchards and vineyards sold for twelve thousand dollars an acre. Incomes, however, did not increase at the same spectacular rates. By one calculation, farms were only producing 2 percent of their market value. In other words, a farm worth half a million dollars would only produce ten thousand dollars in income—about forty thousand dollars less than a similar investment would earn in a bank. Only about a quarter of increased farm wealth came to farmers as income; the rest was a reflection of higher land prices. These paper profits tended to generate unreasonably high expectations. During the boom, equity financing became common. The remorseless rise in farm debt from under $50 billion in 1970 to $190 billion in 1984 followed the rise in land prices as one writer describes it, "like a murderer tracking his victim." Land was the reason why most of the money was borrowed. Lenders looked at soaring land values and rushed to lend, just as farmers looked at tax breaks and low real interest rates and stampeded to borrow, fearing that if they waited too long, they would lose their big chance. In an inflationary cycle it was thought that even the most exorbitant prices could somehow be accommodated.[19]

In the "glory years" of the seventies the seductiveness and persuasiveness of production agriculture—specialization, sophisticated machinery, high-volume production expansion of the land base, and above all reliance on perpetual debt—turned the heads of farm families, and many followed this gospel enthusiastically. Nevertheless, the seventies also saw a revival of concern about rural America, in part as a reaction to big farming. A broad range of groups, from neopopulists to middle-class public-interest organizations to church groups, called for change in agriculture. The emergence of a domestic land reform movement that demanded access to land and the dismantling of the corporate food marketing, processing, and distribution system to meet the needs of consumers was one form of activism. Others centered on environmentalism, seeking to halt the destruction of rural ecosystems. Whatever the nature of the concern, there was general agreement among reformers that the preservation of the family farm, the curbing of corporate agriculture, and the adoption of sustainable farming techniques were essential elements in any blueprint for change.[20]

In California, environmentalists continually hounded big agriculture for its transgressions. A Ralph Nader task force was highly critical of the agribusiness establishment.[21] National Land for the People, the movement to enforce the 160-acre limitation law (which restricted to that acreage those who received irrigation water from federal projects) focused renewed interest on the inequalities in land tenure in the Central Valley and spawned alternatives to industrialized agriculture.[22] The state itself sponsored a task force that investigated the viability of the small farm in California.[23] In the Midwest, more modest but significant activity focused on the implications of corporate ownership for the structure of agriculture. For example, the Iowa Farmers Union, which was instrumental in securing the passage of anticorporate farm ownership legislation in Iowa, helped sponsor a landownership survey in the late seventies as part of its effort to preserve the family farm. The Catholic church, through its rural life departments, also became outspoken on the issues confronting rural America.[24] Much of this activity found a sympathetic ear in the Carter administration, whose secretary of agriculture, Bob Burgland, placed the resources of the United States Department of Agriculture behind an investigation that explored the inequities inherent in the structure of modern agriculture. The valuable final report, *A Time to Choose* (between a system of family farms and one of corporate-run agriculture) resulted from hearings around the country.[25] These promising developments in the reeducation of the agricultural community from the highest levels of government were cut short with the advent of the Reagan administration, which returned to the efficiency theme and sought to reduce the role of government in agriculture.

The Reagan victory sent a clear message to farmers that pointed to the elimination of support programs and a return to the free market so dear to generations of conservative Republicans. The realities of the political economy of agriculture were such, however, that the bold plans of the president's budget director to eliminate subsidies never materialized. "The worst nonsense of all the budget, of course," David Stockman later intoned, "was the farm subsidies. The nation's agriculturalists had never been the same after the New Deal turned the wheat, corn, cotton, and dairy business into a way of life based on organized larceny." Stockman was hoping in March 1981 for a coalition between the southern Democratic Boll Weevils and the midwestern Republicans, which would follow the president's plan to slash government farm programs, but the coalition failed to materialize. "We ended up," wrote Stockman, "signing a new five-year farm bill that amounted to a smorgasbord of everything—production controls, price supports, subsidy payments, export financing. The agriculturalists turned the USDA's welfare handouts and market rigging schemes into awesome surpluses of cheese, wheat, corn, rice, cotton, even mohair and honey."[26]

In the farm crisis of the eighties, the failed "Reagan revolution" played its part, as did a number of other factors, which often transcended the boundaries of individual states and were national, if not international in scope. While there were differences in degree and emphasis and especially in timing, the same forces worked to cause economic stress in the Central Valley as in the corn belt. These forces influenced cotton farmers and corn growers regardless of where they lived.

At the level of the individual farm, the inflation of the seventies played a major role, for one of the primary causes of the crisis involved debt, the emphasis of many farmers on capital gains rather than cash flow. Like their predecessors of 1920, they were left holding expensive mortgages, which declining prices made it impossible to pay. Federal tax and fiscal policies had exacerbated the inflationary climate by pushing ever more investment into land, buildings, and machinery. Unlike other sectors of the economy, agriculture had no defense against inflation. Whereas it became a common practice elsewhere to index wage and price increases, farmers' only defense was to accelerate purchases of capital goods to keep up with inflation. In a deflationary climate, indexing was no longer necessary, but farmers were left with financial commitments that spelled disaster.

At the national level a second important cause was the misconception that worsening world food shortages would create an unlimited demand for United States farm products. The failure to understand that the American economy was no longer self-sufficient but was integrated into the world economy had severe consequences to the balance of payments, which in the seventies depended heavily on huge agricultural exports. Farmers often cited the 1979 Carter trade embargo to Russia as a major reason for the loss of markets. The Federal Reserve Board's decision in 1979-1982 to wring out inflation with a tight money policy also had disadvantages for farmers. Not only did interest rates increase but inflation dropped drastically, leaving farmers with none of the advantages of an inflationary climate and forcing them to service debt with current income. Finally, the Economic Recovery Act of 1981—"the most irresponsible Congressional act in the history of the

Republic," as one agricultural economist called it[27]—which was meant to help farmers make it easier to hand over property to the next generation, actually ensured that massive deficits would be enacted through tax cuts. For farmers, a strong dollar, high interest rates, and falling land values combined to poleax the agricultural economy by lowering exports, boosting the cost of production, and scaring off investors.[28]

Social Movements and Change

It is instructive to compare the agricultural economic crisis of the eighties with that of the twenties in order to obtain a sense of perspective on the forces acting within the agricultural community. Both crises saw a severe economic downturn after a period of rapid inflation, which left some farmers with heavy obligations and without the wherewithal to pay them. Both crisis presented the prospect of continued low incomes for all. In the eighties farmers were a distinct minority, whereas in the twenties, though they were numerically strong, they lacked sufficient political influence to achieve their economic goals. All through the twenties, farmers and their supporters placed their faith in legislative action to solve their problems. Specifically they supported the McNary-Haugen bill in Congress, which aimed at raising farm incomes through a two-price system—one for the domestic market and the other for abroad. A "fair" price would be maintained, first, through a protective tariff and, second, by a government corporation that would buy up enough of each commodity to force up its price to a fair value. Surpluses would then be "dumped" abroad at the prevailing world price. McNary-Haugen passed Congress twice, in 1927 and 1928, only to be vetoed by Coolidge.[29] Farmers, therefore, faced the early 1930s with no protection whatever from market forces. Sixty years later not only was there an institutionalized group of farm programs but, despite the relatively small numbers engaged in farming, the agribusiness lobby was extraordinarily powerful and effective. In 1983, for example, the year of the payment-in-kind program, farmers in Iowa received $925 million in support programs, while those in California received $357 million.[30] The 1985 Food Security Act, described by on USDA official as "easily the most expensive farm bill ever," did not alter these trends, despite the effort of a minority in Congress and outside it to promote mandatory production controls designed to raise farm prices. The Congressional Budget Office estimated that farm spending could run to $56 million in 1987-89, almost as much as Congress spent on all farm subsidies between 1961 and 1981.[31] Because of the poor farm economy, the bill was a kind of holding action with a strategy to subsidize farmers heavily in the short run in the hope that the new policies would wean them of subsidies in the long run. The dairy herd buy-out program, which was designed to cut milk production, illustrates this desire but also the failed logic of the program. For instance, one Kings County, California, dairy farmer was paid over $8 million for his herd, almost as much as was paid for all the herds culled in the four largest Iowa dairy counties.[32] And while millions were spent on the elimination of cows, nothing was done to cut the milk production of dairies that remained.

Each of the various sectors of farming pressed its demands for special treatment and, more often than not, received it. Farm policy is extraordinarily complex, and not surprisingly, public opinion is confused on the issue. Remarkably, the public considers farmers the least influential interest group in our society. Labor unions and business, by contrast, were seen as having too much influence, while physicians, real estate, and the military were not far behind. According to a 1985 Roper poll, the public showed "immense sympathy for honest, hard working, and beleaguered farmers who do society's dirty work, are relatively powerless, and must contend with the uncertainties of weather and markets."[33]

Where did this sympathy for the plight of the farmer originate? After all, in 1930 a quarter of the nation's population was engaged in agriculture, but by 1980 only 3 percent held farm jobs. Apparently, the public feels a sense of familiarity and personal contact with farming, perhaps because, according to a 1985 Gallup poll, almost four out of ten Americans had at some time in their lives lived or worked on a farm, and half had relatives or friends who were farmers. Not surprisingly the public favored government support for farmers in times of difficulty. While they were inclined to believe that farm programs only helped "big" farms and not "family" farms, as many as 36 percent wanted to see federal spending to farmers increased. The public was also aware that only a small percentage of the food dollar went to the farmers; the "middleman" got most of it.[34] Thus agriculture, and especially agribusiness, went into the farm crisis of the 1980s in a unique position: on the one hand, over the years it had built up a powerful and effective lobbying structure in Congress, which ensured that farmers received economic support through flawed and costly government programs; on the other hand, "family farmers" were viewed by the public not as a special interest group but as worthy of government support because of their vulnerability.

In attempting to understand the dynamics of farm activism fifty years ago and today, it is useful to divide such movements into two categories. The first, the "classical" type, has an indigenous base and resources and leadership developed from the aggrieved themselves.

The so called "professional" movement is led by professional reformers, has weak or nonexistent membership, and is mobilized primarily to tap a "conscience constituency," primarily through the electronic news media, which can be relied upon to provide dramatic coverage.[35] Following this kind of analysis, support in earlier crises was mobilized through powerful oratory by such zealots as Sockless Jerry Simpson and Milo Reno.[36] The symbols of protest were unshaven farmers wielding baseball bats at the Sioux City stockyards, or a truckload of striking cotton pickers in Tulare County, California, defiantly waving a banner protesting grower violence. Comment, reportage, and in the 1930s, photographs of these activities were by and large printed in a hostile press. Even when the protesters had an overwhelming case, as when striking cotton pickers were shot down by growers in cold blood at Pixley, California, in 1933, there was little sympathy for their plight except among radical groups.[37] In the contemporary world, television has revolutionized the delivery of news, and as in other areas, farm protesters and their advocates have taken advantage of the medium. The planting of crosses symbolizing foreclosed farms, tractorcades halfway across the country, vigils outside the office of an important official, and other symbolic events are calculated to draw attention to a particular issue and, through television coverage, to spread the message to a viewer audience.

Obviously the climate of crisis in the eighties is very different from that of the twenties and thirties. For example, between 1927 and 1934, there were an estimated 24,600 farm foreclosures in Iowa. In other words 11 percent of the land was lost to mortgage indebtedness in the whole state. Similarly, between 1922 and 1934 there were 7,192 farm bankruptcies in Iowa and 2,575 in California. Once the New Deal programs, especially those of the reorganized Land Bank, were in place after 1936, forbearance and "social lending" policies eased economic stress for the farmer. Until that time, except in isolated instances, they had to rely on their own devices to surmount the consequences of over a decade of hard times.[38] By the eighties rural society no longer operated this way. It was not just the power of sympathetic media, capable of portraying the underdog in a favorable light. Farm families now had the opportunity for support and a fair hearing in any difficulties they encountered. They benefited from the diffusion of concern for individual rights originally initiated by the civil rights movement. Thus in the eighties farmers had reason to expect that they would receive equitable treatment in their dealings with large bureaucratic lending institutions or small country banks. When they did not, assistance was usually near at hand.[39]

Nothing illustrates the change in climate between the eighties and the thirties better than the utilization of popular music and "Farm Aid" concerts to attract money and recruits to a movement promoting alternatives to those championed by production agriculture. In addition, funds were generated for such necessities as legal aid and food pantries and to pay administrative costs for farm advocacy work around the country. By the summer of 1986 some of this activity had become political advocacy at a national level with the convening of the Farmer-Rancher Congress to offer alternatives to the inadequate farm bill of 1985.[40] In the thirties farm activists had none of these support systems and, as the classical model of a protest movement suggests, had to rely on their own membership, resources, and leadership to produce results.[41]

It is important to underline that the farm crisis of the eighties had a selective effect on farm families, whereas the Great Depression stuck in a more general fashion across the board. In the eighties, while one farm family suffered the trauma of financial difficulty, down the road another would be hardly affected by the downturn. Poll data illustrated this phenomenon, and the importance of the 1985 Farm Program in sustaining large numbers of Iowa farm families. In December 1986, for example, as many as 40 percent of all Iowa farmers claimed they were making a comfortable living, while 56 percent were worried about their financial condition. This obvious differentiation in the economic condition of families complicated solutions.[42]

Ironically, despite the more liberal climate of postmodern America, it is capable of showing a more sinister and tragic side than was usual in the more rigid "inner directed" world of the thirties. Circumstances could channel embittered and deranged farmers towards acts of irrational violence against an authority figure in a way that was uncommon fifty years before.[43] At the same time, though there was always the danger that a sheriff's deputy, a loan officer, or a neighbor would become a victim to senseless violence, the mobilization of communities under economic and psychological stress produced an enormous grass-roots effort to promote self-help groups. In this way the chances of desperate acts by members of farm families on themselves and on innocent victims were minimized.

In Iowa this mobilization not only included legal, financial, psychological, and welfare resources for needy farm families, it also brought the residents of small communities together in a movement to stop the decline of their towns and villages suffering from a long-term economic downturn. In California a very different political and economic climate made for another kind of mobilization altogether.

A test of the sociopolitical makeup of agriculture in Iowa and the Central Valley came, therefore, with the periodic economic crises that plagued farming in the twentieth century, especially the 1980s. By comparing the two areas we can explore the different responses to economic stress in agriculture, the complex background of contemporary agricultural conditions, and the past and future place of the farm family. In this book, the first two chapters trace the impact of technological change on the two localities, particularly the replacement of the self-sustaining farm by more specialized farms, whether in livestock, row-crop, or fruit and nut production. The next three, more technical chapters describe how the structure of agriculture affected land tenure, farm inheritance, and credit, mechanisms. I have tried not only to cover ground that has hitherto received little coverage in the historical literature but also to provide a context in which the contemporary debate over the future of farming can take place. Chapters 6 and 7 employ a case-study approach to understand the contribution of the farm family and the farm community to the social structure of rural society. At a time when the economic downturn has made both the local community and the families within it highly vulnerable, it is important to explore the historical record for possible strategies for survival. Finally, the book analyzes the response to and the impact of the farm crisis of the eighties in Iowa and the Central Valley and from this experience draws some conclusions about the future of family farming and its place in agriculture.

1 Corn-Belt Farming

Changes in land tenure, inheritance, and credit mechanisms—subjects to which I will turn shortly—make little impact on the casual observer of agriculture. On the other hand, the abandonment of farmsteads, the introduction of Harvestore silos, the use of huge four-wheel-drive tractors, the wholesale leveling of the land, and vast irrigation projects could not fail to leave some sort of impression that agriculture had indeed changed over the past eighty years. Unquestionably, technological and scientific innovations have altered the way the corn-belt farmer goes about the business of growing crops and raising livestock.

The middle of the eighties finds agriculture in its worst crisis since the thirties. Many of the components of that crisis are due to the very success of farmers at producing basic commodities such as corn, cotton, soybeans, and milk in such abundance that they cannot obtain a fair return. One of the ironies of this success, made possible by highly scientific agriculture, is that it has come at great cost. At the social and cultural level, millions of families have left the land. In the long term, many probably secured better lives for themselves and their families, but certainly many others did not. Modern agricultural production methods, particularly such basic innovations as the application of chemicals and insecticides and the use of irrigation have done much to harm the environment. It would be ridiculous to suggest that the thousands of innovations that have become part of the daily round of farming should be eliminated and the clock turned back. At the same time, it is worth making the point that the present infatuation with technology has its drawbacks. In the corn belt, the overuse of chemicals on the land has polluted the groundwater, and some wells are now unsafe to use. Obviously, at a time when many are calling for a reappraisal of agricultural organization, it is worthwhile to explore how farming evolved from its frugal past to its present condition. First, let us look at the corn belt before the start of the agricultural revolution that began in the forties.

Classic Corn-Belt Agriculture

The genius of the typical corn-belt farm of seventy years ago was the circularity of its operations. The farm grew its own feed, which was then fed to livestock which in turn helped to renew the soil with their manure. Livestock raising was a method of marketing crops that had no other market. What economists call outside inputs—the feed, commercial fertilizer, pesticides, fuel, and even groceries that nowadays must be bought off the farm—were unheard of seventy years ago. Before the introduction of hybrid corn, farms even produced their own seed corn, and although tractors began to replace horses in the early twenties, their diffusion was slowed by the downturn of the economy. As far as the standard of living of the family was concerned, apart from the purchase of a few staples, and a twice-yearly splurge in the Sears or Wards catalogue, store purchases were less common also. Often all the cream, milk, eggs, poultry, and potatoes for the family were supplied directly from the farm, along with substantial amounts of meat.

Nearly all the labor needed on the farm was performed by the farm family. Wives and children under the age of nineteen contributed as much as 15 percent of all labor during the year. The farmer devoted half an hour every day to such household tasks as gardening and fuel hauling. The precise budgeting of a family's time underlines the kind of work ethic and routine that was required sixty years ago. With few labor saving devices either in the home or on the farm, there was little time off for any member of the family.[1]

The Hog-Dairy Enterprise

Nowhere was this kind of dedication and industry better illustrated than in the northeast Iowa counties where dairy and hog farming flourished. Here the frugal farmer reigned supreme. Because the soil was not capable of producing successive grain crops, dairying had been employed since before the turn of the century. Hogs, raised on skim milk and surplus corn, supplemented the dairy. Something in the range of 30 percent of the total farm area was in pasture, with the remainder of the area cropland. Local creameries provided a market for most of the milk produced, and the hogs were shipped to larger markets like Waterloo and Chicago. The average farm size was 150 acres. Almost all the labor, except at peak times, was supplied by the family. In 1923 only 17 percent of all farms had tractors. Four horses were required to work eighty acres of land, and each additional forty acres needed another horse. To emphasize the importance of livestock products on these farms, 42 percent of the total income came from dairy and cattle products, 27 percent from hogs, and 10 percent from poultry. Cash income from crops amounted to only about 5 percent.[2]

The interconnectedness and circularity of the operation, where most farm products were utilized by the farm itself, was the hallmark of the hog-dairy enterprise. In the prechemical era, crop rotation played an important role. About 41 percent of the cropland was in corn, 37 percent in oats, and 25 percent in hay. Corn and oats followed each other in rotation, with every fifth year given over to hay. As a result of good soil management and the use of manure, corn yields on these farms had risen from the turn of the century to a greater extent than in the rest of the state. Moreover, there was a tendency for the higher yields to come from farms with smaller acreages in corn, for the simple reason that these farms concentrated more on dairy production and thus applied more manure to their fields. Farmers tried to be as self-sufficient with their corn as possible. About 18 percent of the corn crop was fed to dairy cattle, and 66 percent went to the hogs. The secret of hog-dairy enterprise was to find the correct mix of production of butter fat and pork from the corn grown. Concentration on hogs and corn, of course, required less labor than dairying. The two other important crops, oats and hay, were used primarily to feed cattle and horses. Much of the hay land was not as productive as it could have been because many wet areas were not drained and were impossible to work in the spring.[3]

During the peak of prosperity, 1915-1920, farmers in the hog-dairy area concentrated on hog production, but in the twenties and through the early Depression before the introduction of price supports, dairy once more came into its own. An average farm kept fourteen sows and seventy-one hogs, producing an average of about thirty thousand pounds of pork per year. In the dairy enterprise a typical farm had fifteen cows, but in 1923 the industry was undergoing a shift from dual-purpose cattle towards speciality dairy animals. The dairy stock was almost all Holstein with a few Jersey or Guernsey herds. As regards the breeding cycle, almost half the herds freshened in the fall months, so that winter dairying could take place when prices were more favorable. About a quarter of all calves were kept to maintain the herd; the rest were fattened on whole milk and sold for veal.[4]

With the onslaught of the Depression, hog prices continued to fall, and for this reason dairying retained its position. Although the crop system remained very similar to that of the twenties, the smaller numbers of hogs allowed farmers to grow more legumes. Soybeans, for instance, were grown for the first time and utilized as a feed for dairy cattle.[5]

In general, no farm families worked harder than those in the hog-dairy enterprise. In the 1930s it was calculated that in northeast Iowa the total labor on a farm was the equivalent of about two men working nine hours per day all the year around. The operator provided half this labor, and the rest of the family furnished approximately 23 percent of it. Hired men were often used from March to October. Farmers usually worked an average of 12½ hours per day from May to November, and even in January they toiled 8½ hours a day. The dairy was the heaviest user of labor throughout the year, especially in the winter, when five or six hours were needed for chores. The hog-dairy combination worked well for the full utilization of labor as well as land. Hogs required relatively less labor but large amounts of corn, while dairy cows needed a great deal of labor but less time spent in forage preparation. According to the detailed figures available, one milk cow required twenty-two minutes a day of labor, a litter of pigs only five minutes. Evidently all this hard work paid off. In spite of the lingering Depression the incomes of hog-dairy farm families were above average.[6]

Diversified Farming

In most of Iowa corn was the staple crop on which farming was built. In one area during the twenties, corn was planted on at least one-third of all land over a period of four years. In the pretractor era and until the early thirties, this corn was the self-pollinating variety with maximum yields of around sixty bushels per acre and an average of around fifty bushels. Continuous corn cropping took a heavy toll on the soil, making a sound rotational system imperative. In the twenties an average of eleven hours of labor was needed to produce an acre of corn up to the time it was picked. Picking by hand required another eight hours per acre. Mechanical corn pickers did not save appreciable amounts of labor. Increased yield was the best way to reduce expenses, and the rapid diffusion of hybrid corn, with its greater yield potential, was therefore all the more understandable.[7]

In corn cropping labor requirements were high during two periods of the growing year. The first, from May to July, took care of planting and crop cultivation. A forty-acre field, on average, needed thirty-six man-hours of labor a week during this time. The second period, from early November to the middle of December, when picking took place, required from twenty to sixty man-hours a week. Counting on an eight-hour day, ten days would be needed to plow forty acres, five days to disk, two and a half to harrow, and three days to plant the corn. All in all, it took about three and a half weeks to fit and plant the seedbed. Later in the summer cultivating would take a further three weeks with a one-row machine. In the fall one man and a team could take care of picking in about thirty-five days, or a month and a half. After that, if the weather held, plowing would take about a week. With corn rotated every two years, manuring was a priority, and about four tons of animal manure was spread on each acre of corn land. Most corn was picked for grain, while much smaller amounts were cut for silage or left in the field to be hogged down. Hogs consumed about 55 percent of the picked corn; about 10 percent was fed to the general-purpose herd of cattle; steers got 13 percent, horses 5 percent, and poultry 3 percent. Only 7 percent was sold on the open market.[8]

Tenant operators increased in the 1920s and during the Depression, and because they paid their rent in shares of the crop, they were forced to devote a larger percentage of the land to corn. The expanded corn growing had a detrimental effect on soil conservation in many parts of Iowa. In the west-central grain area, corn usually took up 40 percent of all acreage and oats about 32 percent. Corn was emphasized most heavily on hog and crop farms, and least on beef-raising operations. Hog farms consumed about 70 percent of the corn they produced. On a rented farm, the size of the hog enterprise was determined by the amount of corn needed for the crop share, but on all farms, before the tumble in prices after 1930, corn that was converted to pork was more valuable than in its natural state. Hogs required relatively little labor except at farrowing time, and their equipment needs were minimal also. Most farmers saw to it that sows farrowed only in the spring, when there was more time to look after the newborn pigs, although those who could find time to raise litters in the fall did so. Farms averaged between six and twelve litters per year.[9]

In Iowa generally the cattle enterprise had four usual forms: the family dairy operation, the dual-purpose herd for both beef and milk, the dairy farm that also produced baby beef, and finally the all-beef herd. Beef raising, which concentrated on breeding and fattening cattle for the market, produced greater income and required greater expenditure than other kinds of operations. In the cash-grain area farmers who specialized in beef raising had to purchase 50 percent of their corn. Their fixed expenses were greater, and in addition, they spent large amounts of money on the purchase of animals for fattening. While hog farms had average expenses of only $2,341, beef farm expenses, in the shape of feed and the cost of animals, totaled $10,561. Although gross incomes were over eight thousand dollars more on beef than hog farms, net incomes were not significantly different.[10]

From 1900 to the late 1930s, the production system in the corn belt remained stable. Corn and oats were the principal crops, with hay and legumes for forage also emphasized. The basic objective of most livestock farms was to remain self-sufficient, so that the crops grown could be recycled into pork, beef, butter, and milk. In some areas of the state, corn was grown for sale on the open market, but in general, whatever the specialization, every farm had a few milk cows, beef cattle, hogs, poultry, and work animals. As the farms relied largely on horse and man power, hours were long, and the family had to work hard. Although the basic orientation of farmers was toward the market, agriculture in the corn belt was also a way of life.

The Revolution

Although the Depression slowed change, the seeds of drastic alteration were planted at that time. The twenty-five-year period from 1928 to 1952 saw a virtual revolution in technology that would irretrievably alter corn belt agriculture. The most obvious changes were the substitution of mechanical power for human and animal energy and the application of scientific ideas to livestock breeding and crop production. With the retirement of horses, most implements were adapted to tractor operations and equipped with rubber tires. Combines replaced binders and stationary threshing machines, and four-row planters and cultivators replaced two-row. The universal availability of electricity was a revolution in itself; water could be pumped with some reliability, and hydraulics allowed many heavy and unpleasant tasks to be performed mechanically. Hybrid corn was rapidly diffused after its introduction in 1928, and after World War II came commercial fertilizer and soybeans. In the twenties most farmers used their own livestock for breeding. By 1952, in some areas of the corn belt 75 percent of them used artificial insemination. Similarly, by the fifties feed had become much more sophisticated, with proteins, concentrates, minerals, vitamins, and antibiotics routinely added to rations. Consequently, about a quarter of all feed had to be bought. Obviously all this change had a drastic impact on the farming operation. At one level, considerable labor time was saved, but the new techniques and machinery cost money and added to the expense of doing business.[11]

This first revolution in agriculture, was in good part the result of government pressure to update and improve family farming devastated by the chaos of the Depression. The USDA, together with state experiment stations, sponsored a vast array of research and hands-on projects to push and pull farming out of the dark ages.

Farmers were slow to adopt the new ideas that came out of such programs in the forties and fifties. For example, it took seven years for the majority of one sample of farmers to adopt weed spraying, with herbicides, and as many as nine years for half to employ antibiotics in hog production.[12] Another study traced the adoption of commercial fertilizer in corn. Again, farmers were conservative, and only 41 percent of all corn acreage was given an application in 1953. The slowness of diffusion was in marked contrast to the rapid change that would occur in the sixties and seventies.[13]

The fifties were also a decade when large tracts of land that had previously been unsuitable for cultivation were brought into production. The quest for a more rational agriculture, with larger farms and bigger machines, saw a wholesale assault on land that was too wet to till. Some drainage had been attempted in Iowa in the nineteenth century, but its expense prevented adoption on marginal land until after World War II, when government programs began assisting farmers.

The purpose of drainage was to control soil moisture by removing excess water from the upper three or four feet of soil. On well-drained soil, water drained downward to a water table more than four feet below the ground. But on poorly drained land, the water remained near the surface, preventing work in the spring. Tiling and open drainage systems solved the problem of standing water, allowing a longer growing season, better use of all available land, and more productive farms.

One of the more positive legacies of the Depression was a concern for the conservation of resources and the scientific application of methods in this direction. Appropriate land use and fertilization practices needed to be supplemented by supporting practices such as contour plowing and sod waterways if soil was to be conserved and improved. Contour cultivation, strip cropping, and terracing were comparatively simple measures designed to prevent water runoff and soil erosion. Any of these measures could be used by itself, but strip cropping and terracing were usually introduced to blend into the contours of the land. All these practices were profitable, since they permitted more intensive cultivation, reduced loss of topsoil, and increased crop yields. On more level ground the introduction of grass waterways had the same effect as terracing in that the waterways assisted water runoff without damage to the land.[14]

Together with these specific attempts to improve productivity, there were also efforts to change the structure of land tenure. To achieve this "adjustment" of agriculture, as it was called, those on "inadequate-sized" farms were encouraged to seek nonfarm occupation so that those who preferred to stay on the land could obtain sufficient property to farm more efficiently. Those who chose not to expand but to stay on the farm were also urged to seek nonfarm employment. Although nothing like an official policy dictated that these ideas be carried out, family farms in fact disappeared on a massive scale in the forties and fifties, effectively achieving the informal objective through voluntary migration. A good measure of how far farm adjustment had progressed was given by a 1954 Iowa study, which showed that 28 percent of all farms in the state could not provide a family a decent living. The resources on the average farm were out of balance, largely because there was too little land and capital available to make use of the supply of labor and machinery. According to the 1954 census of agriculture, the average Iowa farm was 180 acres, of which 140 acres were in production. The typical land-use pattern had 55 acres in corn, 11 acres in soybeans, 31 acres in oats, and the remainder rotated in hay and pasture. The average farm had 5 dairy and 5 beef cows, 12 feeder cattle, 3 ewes, 13 litters of pigs, 120 hens, and 1.5 tractors. The available labor supply consisted of the equivalent of 1.4 workers, who toiled three thousand hours annually. Using reasonable labor requirments for growing and harvesting crops and producing livestock, the actual labor required for the average operator was only just over two thousand hours. Thus there was nearly 50 percent more labor available than was required at an acceptable level of efficiency on the average farm.[15]

The message from these findings was that a further "adjustment" was needed, which entailed either an increase in the size of the farm or a transfer of excess labor capacity to off-farm employment. The first alternative would require a large increase in crop acreage in order to fully utilize the time of the farmer. Thus, corn-belt agriculture was on a course that would produce ever-increasing farm size, the utilization of big machinery, and a reorientation towards grain farming. All these strategies would be adopted in the sixties and seventies and would lead to what has been called the second revolution in agriculture and eventually to the disastrous downturn in land and commodity prices that has characterized the eighties.

Modern Corn-Belt Agriculture

In the fifties a typical corn-belt operation still practiced a strategy of risk aversion. It was still more profitable to feed grain to livestock than to try and sell it on the open market. Usually hogs and dairy were combined in areas where the land was of mediocre quality, as in northeastern Iowa; hogs and beef were emphasized on the rich soils of the central and western areas of the state. Some farms still combined all three. In every case, however, the strategy was to stay flexible enough to take advantage of price changes. Farmers watched the markets and planned crops and livestock accordingly.

Changes in crop production technology in the field and in nutrition and health technology in the barn had a profound effect on this flexibility. In many cases the technological breakthroughs cleared the way for specialization in grain farming, pork and beef production, or dairy. Farming began to require more specialized technical knowledge, and the new equipment was not very adaptable. Four-wheel-drive tractors, for example, were unsuited for multiple use in both crop and livestock enterprises, and so many farmers who purchased such tractors dropped hogs to concentrate entirely on grain production. New crop technology also permitted intensive production of corn and other grains on land that had hitherto required a rotation of crops. As the value of the grain produced on this land for sale at the market came to exceed the value that could be obtained by using the land for grazing or for growing feed crops, many families, especially those with more productive land, became grain farmers. Then, in the 1950s midwestern dairy farmers had to decide whether to upgrade their standards of production from Grade B to Grade A or to leave dairy farming. Farmers who chose to upgrade had to purchase a bulk refrigeration tank in addition to pipelines to move the milk from the milking parlor to the dairy. This was a relatively expensive investment—$3,900 for a five-hundred-gallon tank in 1958—more than many farmers could afford. And so the fifties were a watershed in the history of Iowa dairy farming. Although many updated their facilities, others did not; these either retired or devoted themselves entirely to hogs or grain production.[16]

Specialization in corn and soybean production had a major effect on the way the average farm was run. Farmers began concentrating the year into three months of hectic work. Because of the variablity of weather conditions, planting and harvesting needed to be done at the first possible opportunity. Hence, larger machinery and the maximization of available acreage through the application of chemical fertilizers, herbicides, and insecticides became necessary. However, the movement towards specialization in grain farming was gradual. As late as 1976, only 13 percent of all Iowa farmers reported grain as their only produce, while as many as 85 percent raised some kind of livestock.[17]

On any type of farm it was important to select the appropriate machinery for tillage, planting, weed control, and harvesting. Large machinery reduced labor costs but machines had higher interest and fixed costs. Fuel price increases were especially bothersome when energy prices soared after 1973. As might be expected, there was a broad range of machinery in use. Iowa farms had an average of three tractors in 1976 of which 63 percent were under eighty horsepower, although most of the newer tractors were larger models. As was true in the fifties, smaller farms often had more machinery than was necessary for efficiency. The fuel crisis also sparked interest in reduced tillage, with the moldboard plow becoming increasingly redundant. By the eighties no-till implements were fairly common.[18]

From the middle sixties onwards, net income in beef raising and feeding operations lagged behind other enterprises. The traditional advantages Iowa cattle feeders had possessed in earlier decades were lost to large-scale western feedlots, many of which were financed as tax write-offs by business people from outside agriculture. In 1976, 21 percent of all Iowa farms reported selling cattle as feeders, while 37 percent sold fed cattle for slaughter. Numbers of animals were generally small—an average of 55 feeders and 108 fed cattle per farm.[19] Beef raising went through particularly hard times after 1975, when returns on investment only once went above 7 percent.[20] The poor performance of the Iowa cattle enterprise was symptomatic of the management practices of the thousands of small operators in the state. Acting as individuals in a notoriously risky environment, many were not skilled enough in modern management practices, such as hedging, to be able to compete with the huge volume of their western competition. Often they learned too late that diversification was the only way of remaining afloat.[21]

Hog farming alone retained its former importance. But here again structural change caused significant alterations. By 1950 a shift from pasture to drylot raising had begun, prompted by a number of factors, including the introduction of better nutrition, partial elimination of dangers to health from parasites, availability of environmentally controlled housing, introduction of waste-managment techniques, and mechanization.[22]

Technology was the driving force that allowed hog farming to shift from small manually operated enterprises to larger ones, often in more industrialized settings. The opportunity to confine all animals except sows about to farrow permitted a year-round factorylike operation from farrow to feeder to finish. Because of advanced nutrition and the control of disease and parasites, hogs no longer needed grazing time in the field to consume essential proteins and minerals, nor did they need to be rotated from site to site to remain healthy. Although it was technically possible for hogs to be produced separate from a land base, most hogs continued to be raised on medium-sized family farms: 83 percent of farms with sales over $2,500 farrowed pigs in 1976. These farms not only produced most of the grain needed in the operation, they usually had the equity and earnings capacity to invest in the new kind of hog facilities.

The changes in hog production on the corn-belt farm provide a case study in the way credit was employed for expansion and how government farm policy and programs favored technological change in the late sixties and seventies. By then, the typical hog operation could no longer rely on a production system geared to a few A-frames out in the pasture but required, it possible, an environmentally controlled building and mechanization to handle feed, water, waste, heat, and air. Such facilities were expensive, and even modest designs required borrowing quite substantial sums of money. Ironically, this need for financing was welcomed by observers, who claimed that it would make farmers more proficient managers and would force country banks to adjust to find funds to retain hog-producing customers. Government tax policy also encouraged the adoption of the new technology in hog production, for investments in durable production assets could be written off at an accelerated rate of depreciation during the early years of operation. Moreover, investment tax credits reduced the income tax due by a percentage of the initial investment. Since much of this expansion took place during a time of inflation, it became important for a operator to introduce these capital-intensive technologies as rapidly as possible before further price rises canceled their impact.[23]

The hog production confinement unit capable of handling over five thousand animals a year and capitalized at well over half a million dollars was often a prime candidate for a new kind of farm organization, the Sub Chapter S Corporation, in which the income to the corporation passed directly to the shareholders rather than to the corporate entity itself. In this kind of corporation several neighborhood farmers would become partners. These were set up in the middle seventies, often with the assistance of seed companies, which saw the opportunities for supplying a whole variety of specialized feeds to confinement units. Unlike owners of western feedlots, most partners were family farmers, and typically such units provided the partners with a source of healthy feeder pigs, as well as freeing them from the intensive duties required in farrowing. Partners could concentrate on their own operations while a hired manager took care of the facility, which was under one roof. The cycle of production continued year-round with almost mechanical precision. In a 150-sow operation, with 50 of them farrowing every other month, 30 days would be devoted to prebreeding, 112 to farrowing, and 42 days to weaning pigs. The sows would then begin the cycle all over again, and their offspring would move to the finishing facility, where they would be marketed at around 235 pounds roughly two hundred days after birth. On the face of it, such a system would seem ideal for exploitation by corporate interests. In Iowa, however, anticorporate farming statutes, the specialized nature of the tasks involved, the relatively low pay, the unpleasantness of the daily round, and the lack of qualified personnel preserved hog raising for the family-managed family-run operation. Hog production, thus, did not go the way of poultry, where vertical integration by corporations eliminated participation in production by family farmers. Nevertheless, the industry became increasingly specialized, and in the eighties custom feeding began to be introduced by corporate interests. Here, a farmer was paid a set fee for feeding pigs until they were ready for market. Such procedures were only possible because massive dosages of antibiotics eliminated the disease factor.[24] Between 1975 and 1984, hog raising remained by far the most profitable kind of farming in Iowa, providing a net average income of $17,540 per enterprise, compared to $13,179 for a cash-grain farm. Hog raising was also more profitable in terms of return on investment, especially in the years 1973 and 1975-1978. In 1984 the typical Iowa farm specializing in pork production raised 128 litters of pigs, whereas production on less-specialized farms was only 50 litters, with 535 animals sold at market.[25]

The routine demanded of milk production requires a special dedication even under modern conditions. Unlike California dairies, those in the Midwest depend mostly on family labor. Even in an two-generation business, with children available to do chores as well, it is only feasible to milk a herd of around sixty-five cows and in 1984 the average Iowa herd was only thirty-five animals.[26] Inclement winter weather often forces farmers to keep cows in the barn for long periods, adding to the daily chore time. Even so, many farms still retain other enterprises, and the hog-dairy operation has surprising strength in northeast Iowa.

The dairy business has gone through massive changes in the north-central states since the 1940s. Consumer tastes changed, government regulations eliminated old-fashioned production methods, and although the numbers of cows were reduced drastically—in Iowa there was a 43 percent decrease in the number of cows on farms from 1940 to the middle sixties—biotechnological breakthroughs in breeding enabled animals to produce ever greater quantities of milk. The butterfat content of milk was greatly reduced, as consumer taste shifted to nonfat products. In the late seventies one of the few bright spots for milk production was the rise of the pizza industry, which required large amounts of cheese.

Because Iowa was slower than some states to mandate Grade A conditions in dairies specialization took longer to occur. For those Iowa farmers who stayed in dairying after the fifties, economies of scale dictated larger herds. Quite simply, labor requirements per cow declined as herd sizes increased; likewise, fixed overhead costs fell as the number of cows rose. Since dairying was only a minor part of Iowa agriculture from the sixties onwards, the trends towards larger herds and far greater production had small impact, but nationally they were of major importance, particularly in regard to dairy price supports.[27]

Given the structural changes in Iowa agriculture in the sixties and seventies, many families preferred to leave dairying. To be sure, the dairy herd fitted well into a diversified farm program in some of the rougher country in northeast Iowa. Cows are efficient consumers of roughage, and they maintain soil fertility, as well as providing cash flow throughout the year. Yet the very high labor and relatively high capital requirements of dairying are considered disadvantageous by modern farmers, who typically want a quick return for their efforts. Furthermore there are many health hazards associated with dairy cows.

Iowa dairy farmers who remained in farming possess a discernible ethic and a dedication to the way of life necessitated by dairying. Cows must be milked every twelve hours, and some particularly conscientious families milk three times every twenty-four hours for better efficiency. To maintain such a schedule family members must play a major role in the operation. Most dairymen's wives assist with milking and raise calves, and even with the large array of equipment associated with a modern dairy operation, work is hectic, especially in summer. The need for storage space, particularly silos and the expensive Harvestores, as well as labor-saving machinery—loaders and conveyers for manure and hay-making and silage equipment, in addition to the usual complement of farm machinery—make dairying a highly captialized business. Moreover, dairies use far more energy than other farms to run all this machinery along with milking machines. Machine and power costs are $102 per rotated acre on a dairy farm, compared to $45 on a grain farm. Veterinarian and breeding fees, as well as the expense associated with producing breeding stock, also lower dairy incomes. In 1984 dairying in Iowa had a lower return on investment than did beef feeding, and net income was under half that of hog raising. Enterprises that specialized only in dairying had the highest debt-to-asset ratios in Iowa farming in 1984.[28]

The prognosis for dairying in Iowa is not promising. As the embodiment of all that family farming stood for, the northeast Iowa dairy farmers need a certain resourcefulness to survive the boom and bust economy of the late seventies and eighties and the uncertain future of the federal support programs.

Conclusion

Compared to that in some other states, the modernization of Iowa agriculture was comparatively slow, because many farmers viewed innovations with skepticism. Gradually, however, powerful forces in the agribusiness establishment—the Extension Service, the grain trade, seed companies, machinery manufacturers, and lenders—made their influence felt and encouraged farmers to enlarge their operations and specialize. While a surprisingly large number of Iowa farmers still raised some livestock in the middle seventies, grain production for the market had an increasing appeal. Not only did this kind of farming provide substantial returns, it also relieved the farmer of the kind of drudgery associated with animal husbandry. Yet this dependence on grain production at the expense of diversification would return to haunt both farm families and agribusiness in the eighties.

2 Central Valley Ranching

It is difficult now to appreciate the fairly primitive conditions under which farmers toiled seventy years ago in the Central Valley. Although the climate was exotic and in many areas close to the rivers both ditch and groundwater was plentiful for irrigation, most farm families approached making a living from the soil with a perspective derived from the more humid East. Until the early forties a large number of family-sized ranches were mixed farming operations with a livestock emphasis. Many relied on the high water table to provide water for the irrigation of alfalfa, which became a staple feed for small dairy, cattle, and hog ranches. It was not surprising, then, that a district north of Hanford in Kings County was named Lucerne. Here, farmers could rely on water seepage to grow alfalfa, often producing half a dozen crops a year. Indeed, the ease with which irrigated crops like alfalfa could be grown allowed wasteful irrigation practices to develop that would take years to eliminate.[1]

But if some transplanted midwesterners felt more comfortable re-creating Iowa with palm trees, others quickly adapted to specialization, especially in dairy and fruit. Away from the rivers, large bonanza operators rented huge tracts on which they grew nonirrigated winter wheat and barley. In the early twentieth century in Kings County, the Tulare Lake bottom proved ideal for this type of operation and attracted growers willing to gamble, to bet against the market and the weather. This kind of rancher was the direct descendent of the nineteenth-century wheat barons made notorious by Frank Norris, and this style would be developed and perfected in the thirties and forties by the cotton barons in Corcoran.[2]

These generalizations are supported by the material in table 1 that shows the size and crop production of ranches in Kings County just after the turn of the century. Small owner operators were dominant; only 20 percent of the farms were rented. Most ranches were between 40 and 240 acres, and fruit, grapes, alfalfa, and grain were grown on many of these small farms, usually on tracts of less than 20 acres. As

Table 1. Ranch Size and Crop Acreage, Kings County, California, 1901

			Size of Tracts on All Ranches Devoted to					
Acres	Total Ranch		Fruit/Grapes		Alfalfa		Grain	
per Ranch	%	(N)	%	(N)	%	(N)	%	(N)
<20	17	(126)	67	(339)	48	(271)	28	(111)
20-39	14	(103)	16	(82)	19	(102)	19	(74)
40-79	20	(146)	10	(50)	15	(83)	23	(91)
80-159	20	(147)	4	(20)	10	(57)	19	(74)
160-239	15	(111)	1	(6)	5	(27)	6	(25)
240-319	4	(28)	1	(6)	2	(9)	2	(9)
320-639	7	(52)	1	(3)	1	(5)	3	(10)
>640	2	(15)	—	(1)	—	—	1	(3)
Total	99	(72)	100	(507)	100	(559)	101	(397)

Source: Kings County Directory (Hanford, Calif.: *Hanford Daily Journal*, 1901), 81-144.

Note: The Miller and Lux organization (the largest California farm operation at that time) rented 33,405 acres in the county; these are excluded from the analysis. Percentage figures have been rounded.

many as 70 percent of all operations had vines or fruit trees. In addition, 77 percent grew alfalfa (those that grew alfalfa presumably ran a dairy or raised livestock), and 55 percent grew grain. Among the 13 percent of ranches over 240 acres, eight were over 1,000 acres in size—the largest being 8,229 acres. Three of these operations were dry grain farms, and a fourth was a stock ranch. All were in the southern part of the county, away from the Kings River. It was the small dairy or fruit ranch, however, that dominated the northern third of Kings County and gave this area its special character in the years before the Depression.

Small-Scale Ranching

It was possible to make a living on a forty-acre dairy ranch in the Central Valley sixty years ago. Such an operation usually had about twenty cows producing whole milk for processing and would be classified nowadays as a Grade B dairy. Ranchers grew all their feed, ideally practicing a nine-year rotation—alfalfa for six years, followed by three years of double-cropping with barley and vetch in the winter and Sudan grass in the summer. The work load, which revolved around planting and caring for the various crops, as well as managing the dairy herd, kept the operator busy throughout the year. The daily routine consisted of six hours in the field and a further six hours working with the cows. On a totally unmechanized farm, 88 percent of the operator's time was taken up with routine labor using manpower and horsepower.[3]

In 1923 the California Agricultural Experiment Station conducted a detailed study of dairying in California, including the region south of Fresno. There, specialized operations were larger than the average midwestern dairy. In the sample, herds ranged from 16 cows to 103. Although many ranches had fruit orchards and were mostly self-supporting in foodstuffs, essentially they were commercial dairies, with all their productive capacity geared to selling milk or milk products. Homegrown alfalfa and pasture provided the bulk of the feed for the herd, but some ranches supplemented this feed with concentrates that were exotic by midwestern standards—coconut meal, beet pulp, molasses, and cottonseed meal, for example. The prevailing good weather enabled ranchers to keep their cows out for much of the year. Indeed, most used pasture all the year round; they reserved irrigated alfalfa for milking cows and fed native salt grass to dry stock. As in the Midwest, the preferred breed was the Holstein, with as much as 10 percent of the herd purebred. Breeding could take place throughout the year because the weather, labor conditions, and feed made the maintenance of a full milking herd perfectly feasible. Milking was done on most dairies at twelve-hour intervals, at three o'clock in the morning and three in the afternoon. Many employed milkers to assist in the daily routine, and almost half the sample in 1923 used milking machines, but these represented virtually the only mechanization. Most sold their milk to a local creamery for the eventual manufacture of butter. In this, and in many other respects, California dairying resembled that found in the Midwest. Many Central Valley dairies were family-run enterprises that required the same demanding regimen Iowans followed on their farms. Yet by 1923 larger operations already possessed a commercial emphasis that would develop over the years into a highly professional approach to the dairy business, with great care taken in breeding and in the maintenance of the herd and of the land.[4]

Fruit growing bore less resemblance to Iowa farming, for commercial orchards were rare in the corn belt. Not so in California. From the turn of the century to the thirties, the Kings River delta was a major center of soft-fruit production. All the fruit, except cling peaches, was "three-way" marketable (that is, salable dried, fresh, or canned), but in fact, because of the relative difficulty of marketing soft fresh fruit, almost all the crop was either canned or dried. Families could make a living on as little as twenty acres, but the most common size for a ranch was forty acres—typically eight acres of canning apricots, ten acres of black juice grapes, and twenty acres of canning peaches of different varieties. The yearly routine on such an operation began in November with the pruning of peaches and apricots, followed by grape pruning in January. All fruit trees were sprayed in February and March, and the first irrigation for the larger fruit took place in April, when the grapes were dusted and the ground around them disked and hoed. By May the grapes required their first irrigation, and they were also cultivated. The peaches needed to be thinned, and the apricots supported with props. Cultivating and irrigating continued in June, with the apricot harvest beginning late in the month. Throughout July and August both the apricots and peaches were picked by a crew brought in by a contractor; the juice grapes were harvested in late September and early October.[5]

One of the features of this kind of operation was that it not only provided a tight-packed schedule for the rancher but required considerable outside labor as well. A total of 875 man-days annually were required to perform all the work on the farm, but only 205 were performed by the rancher. The operator's tasks included all the cultivation, the cleaning of irrigation ditches, the operation and servicing of the small tractor, part of the irrigation and pruning, and supervision of the harvest. While such farms had no permanent employees, pruners, sprayers, fruit thinners, irrigators, and a crew of pickers were required at various times in the year. Unlike the crop farmer, whose family could fill temporary extra labor requirements, almost a quarter of fruit farmers' expenses were labor costs. All in all, the forty-acre fruit farm was a comparatively complex operation. Unlike the diversified farm, it required considerable management skills.[6]

Of all the California farming enterprises, soft fruit and grapes probably changed least over the years. Soft fruit still has to be picked by hand. And one of the most venerable techniques, the drying of fruit, is still carried out in much the same way as it was sixty years ago. Dried apricots, figs, peaches, raisins, and especially plums in the form of prunes make up a smaller part of the market than canned fruit. However, they have an advantage in more flexible marketing arrangements. Unlike fresh fruit, whether it is to be sold to canneries or to stores, dried fruit can be held back for a favorable price. Moreover, apricots and peaches to be sold as fresh must be picked when the fruit has a uniform color and size, and so, orchards must be picked several times. Fruit marketed through the drying process, especially prunes, can simply be knocked from the trees.

Generally, in the twenties, raisin grapes were harvested at one picking, cut from their stems with a sharp knife, and spread in bunches directly onto trays beside the vines. The trays were then placed on a ridge of soil, thrown up in such a way as to catch a maximum amount of sun. After about four days, the trays were turned to ensure uniform drying. Once all the raisins had turned brown and were two-thirds dry, the trays were stacked on top of each other. Paper trays were used only for rapidly drying seedless raisins. In this instance, when the raisins were nearly dry and ready to be stacked, the papers were rolled up like logs of wood. Once the raisins were sufficiently dry to be rolled between the fingers without exuding moisture, they were emptied into sweat boxes in the vineyard. After another three weeks and a thorough equalization of moisture, they were delivered to the packinghouse. Fortunately, harvesttime in the Central Valley was usually dry, for a period of wet weather could be disastrous. Even a brief shower created considerable extra labor. In the event of rain, all trays had to be stacked or rolled up to guard against damage.[7]

Techniques for apricots, peaches, and prunes were somewhat different. Unlike grapes, they were removed from the orchard for washing, sulphuring, and in the case of prunes, dipping. Both apricots and peaches were cut with a sharp knife, the pit was removed, and then they were laid out on trays with the cut surface up. Prunes were dipped in a hot alkali solution as soon after harvesting as possible in order to shrivel their skins. They were then placed on trays in the drying yard. Apricots and peaches were exposed to sulphur fumes for four or five hours before drying in order to make their skins look transluscent rather than raw. They were then transferred to the drying yard, where they were exposed to the sun for as much as five days. Finally, they were placed in trays stacked one on top of each other for roughly a week. Prunes were treated in similar fashion and also stirred to prevent them from sticking together.[8]

Fruit growing had, and still has, a number of compensations lacking in dairying or row-crop farming. The trees are beautiful in spring, and there is great satisfaction in producing a high-grade product that is pleasing to the eye and taste. These pleasures have remained characteristic of fruit culture, but changes in the structure of agriculture have altered the relatively stress-free environment in which fruit was grown for canning and drying. Processed fruit has lost favor in the marketplace, closing local canneries and packinghouses and dissolving the traditional local interdependence of rancher and town dweller. Ranchers must now operate almost entirely in the risky fresh fruit market.

The Hydraulic Society

Very little would grow in the Central Valley without irrigation, and no discussion of the development of agriculture could ignore the hydraulic society as it grew from modest beginnings. It has been suggested that the year 1931 represents a dividing line in California agricultural history. By then most Californians no longer believed that the health of their economy depended on the existence of the family farm or that water policy should be used as an "agent to transform society," to create a carbon copy of the midwestern life-style. After that year irrigation policy became an ally of the agricultural establishment.[9] From the perspective of the Kings River area, such an analysis would seem to have some merit. Broadly speaking, the years before the Depression were a period of local involvement in the delivery of water. On the south side of the Kings River, pioneer farmers themselves brought the water to the land. They began laying out primitive irrigation ditches and incorporating mutual canal companies in which they took shares in stock. Unlike the many colonies developed by private capitalists in Fresno County with funding raised from lending institutions, Kings County development came through the efforts of the water users themselves.[10]

A share in a mutual water company represented not only a proportional part of ownership but also the right to receive water and an obligation to pay for ditches and to keep them in repair. From the early days farmers could pay for their assessments through labor instead of cash. In this very flexible and "democratic" system, farmers could purchase as little as 1/600 of a share to irrigate one acre. Shares were not tied to the land. They could be transferred from one tract to another, rented to other landowners for a season, or even sold to any willing buyer, and this feature would have a direct impact on the small farmers of the Delta region after World War II. Such flexibility was also translated into the operations of the various water companies. Unlike the colony systems in Fresno, there was no scheduling of water delivery; water was delivered on demand. Ranchers received water on a pro rata basis according to how much stock they owned, and water could be delivered in any location. Shareholders could take as much water as they needed when water was plentiful and there was little policing done by ditch tenders. Even in times of drought, operators were trusted to take water only during the time period allotted them.[11]

Two methods of irrigation were favored. The first involved flooding part of a field, with checks at five-acre intervals. The water was allowed to spread over the field to a certain level and left to stand. The second employed furrows. Water would run down five furrows at a time, so that gradually a whole field was covered. While such a system had its attractions as far as simplicity of method and organization was concerned, it was very inefficient.[12]

The Peoples Ditch company was one of the oldest in the area. In 1918 it had sixty miles of canals, of which the company operated and maintained thirty-three miles, with a further forty miles of laterals looked after by individual farmers. The affairs of the company were managed by an elected board. Because the area was dairy country, stock water had priority over irrigation. Throughout the year, each lateral had to have sufficient water running in it so that a rancher at the far end of the ditch could water stock at any time. No records were kept of the acreage irrigated, and no formal written demands were made for delivery; forecasting was therefore impossible. The natural cycle of the river encouraged inefficiency and waste. Because the season of plentiful water was short—from February to June—farmers were strongly inclined to put all they could on the ground while they had the opportunity. As a result water was used lavishly in order to force it over the ground in a hurry.[13]

The Peoples Ditch was located so that even in a dry year it received some river water. Other companies farther downriver could not anticipate such regular deliveries, however. Only with the construction of a reservoir could the flow of water be stabilized and rationed. Ironically, the lavish use of water eventually produced subirrigation. Water percolated into the water table and raised it considerably, rendering surface irrigation unnecessary. Unfortunately, subirrigation also caused alkali buildup in the soil, which took thousands of acres out of production.[14]

The Revolution

Because California was in the forefront of so many innovations in agriculture, it is less easy to pinpoint the period when farming and ranching "took off" on the road to modernization than it is in the Midwest. As a pioneer in marketing, agricultural finance, large-scale farming, and the diffusion of the latest field techniques by its Experiment Station staff, the state led the rest of the nation in many areas. In the southern San Joaquin, however, the introduction of cotton culture seems to have been a turning point. Agricultural finance was profoundly affected, as were labor, farm mechanization, crop rotation, and field techniques. The introduction of cotton to California also involved the federal government at least a decade before the Depression, when similar kinds of intrustions became common.

Government sponsorship at both the federal and state levels had important implications for the diffusion of cotton culture and its success in California. The founding of the cotton research station at Shafter in Kern County in 1919 demonstrated the desire of the United States Department of Agriculture to support the irrigation of cotton in a nonhumid environment. State support came with the passage of a pure seed program, which restricted growers to a single, medium-staple variety of cotton, Alcala. The resultant uniformity gave California cotton a decided advantage over lower-quality southern cotton in the marketplace.[15]

The Central Valley, with its dry, warm climate, its two-hundred-day growing season, and its fertile soil, was ideal for growing cotton. Much of the land was level or could be leveled to irrigation specifications. Rain usually held off during harvesttime, and the warm weather during the fall enabled the crop to ripen uniformly, making repeated pickings unnecessary. Moreover, cotton was not especially susceptible to alkali, which tended to build up in irrigated soil. When it was planted in rotation with alfalfa, yields could be maintained with fertilization and adequate but not excessive irrigation. Alfalfa resupplied the soil with much-needed nitrogen. From an economic viewpoint, cotton provided cash to farmers who had either grown alfalfa as a cash crop or fed it to their own livestock. The income from cotton allowed them to level their land and to add other improvements as well. Fruit growers were less eager to switch to cotton, however, for to do so was to abandon trees or vines that might bear for many more years. Often this decision was made easier in the thirties because the low prices for fruit made cotton seem an attractive alternative until government quotas restricted planting between 1934 and 1936. Thanks to government aid and credit backing from the ginners, by the late thirties, the diffusion of cotton culture was complete.[16]

Cotton was the catalyst for many changes in agriculture between the wars. Not the least important was the new need for talented managers and an army of unskilled laborers. When cotton was introduced in the Central Valley, one promoter seemed under no illusions about the outcome: "The social question involved with the introduction of cotton culture is a serious issue. No one wants an industry in California which is going to further involve the labor problems of the state. Cotton requires large numbers of laborers for thinning and harvesting, and unless this can be handled by having amounts of cotton in rotation with other crops, or be handled by the labor normally available, the social results will not be good."[17] Other ripple effects involved with the introduction of irrigated cotton were the eventual initiation of a statewide irrigation policy and the reclamation of huge tracts of desert land for production. Usually reclamation was achieved by developers, who were charged a nominal rent for a substantial period by landlords in return for the transformation of the property. Here the big Kings County ranchers would play a vital role.

The cycle of production of irrigated cotton usually began in the winter and early spring with plowing, followed by the leveling of the land. Before planting, the soil had to be preirrigated, using either a "flat" or a furrow technique. The former, used primarily with larger fields, utilized flat ground and small levees thrown up at intervals to channel water along the slope of the land. The furrow technique, called listing, threw up ridges about six inches high at forty-inch intervals. During preirrigation, water flowed between the ridges until one acre-foot of water (i.e., the equivalent of one foot of water over one acre of space) covered the field.

Planting usually took place as soon after April 1 as possible, with plants sprouting from the ground five days to two weeks afterwards. Early cultivation was essential to keep weeds from overwhelming the young plants, and later cultivations took place at regular intervals before harvest. The next major task was cotton chopping. In May and June a crew would work through the field, thinning the plants. On smaller farms the crew consisted of relatives, while on larger spreads an army of outside labor did the job. Once cotton began to grow, it needed just the right amount of water to sustain growth, as well as to protect it from disease. Cotton requires about thirty-five to forty inches of water per acre per season to make up for transpiration and evaporation. Severe water stress could cause leaf drop or the premature cracking and opening of cotton bolls. Yet too much water could cause excessive vegetative growth and lower yields. Although the boll weevil was never a threat in California, there were many other pests to defeat, and the industry was quick to utilize chemicals to keep them in check. After World War II, DDT was massively applied, using another innovation, the crop duster. By the first or second week in October, cotton was usually ready to pick. Hand picking could last from October all the way through the following spring, depending on the weather, but most of the crop had usually been hauled in trailers to the local gin by February. There, it was processed into bales for shipment to buyers both overseas and in the United States.[18]

No other community assisted in the diffusion of the cotton culture in California to a greater degree than Corcoran. The social and political impact of the large lake bottom farms in and around this small Kings County town will be discussed in a later chapter. We need only recognize their importance as the generators of change in a very important segment of the economy of both Kings County and the southern San Joaquin. In water procurement and management, in row-crop agronomy, in the adoption and application of complex machinery to farming, they were innovators whose methods were noted and copied by others.

The transformation from the old hit-or-miss bonanza grain farming, which was the style at the big operators in the first two decades of the century, to something approximating contemporary organization, was attributable in part to those who began their careers around the time of World War I. No one typifies this style of farming better than the legendary Elmer Von Glahn. Von Glahn was brought up in the around the agricultural center of Stockton, where he became familiar with large harvesting machinery and pumping equipment. All his farming career he remained more comfortable with grain production and had only a peripheral interest in cotton. It was his technological expertise and especially his organizational flair that were duplicated by cotton farmers. Von Glahn moved south down the San Joaquin around 1915 and began renting land on the lake bottom. Even then he used big machinery in his operation and quickly made a reputation for the adaptation of implements to lake-bottom conditions. Von Glahn was also a gambler; once in the early twenties and again in the early thirties, he lost everything trying to corner the wheat and barley markets. It was in 1934, after working for a few years for another large operator, that he was at last able to put together the kind of organization his talents warranted. He bought some foreclosed land with a $400,000 installment earned from his rental wheat business. By the middle of the war he was farming fifty-five thousand acres of land in one block south of Corcoran.[19]

Von Glahn's mechanical flair was also useful in irrigation and flood control. The late thirties coincided with a cycle of wet years, during which large sections of the lake bottom were under water. For purposes of both flood control and irrigation, Von Glahn dredged a huge main canal, broken into twelve sections. At the upper end of each section was a forty-two-inch centrifugal pump capable of lifting thirty thousand gallons a minute over the dam into the next section. The pumps worked night and day, pushing water through the various stages and onto the fields where it was needed for irrigation. For fighting floods Von Glahn designed a huge floating dredge with a three-yard bucket. This could be rushed to a danger spot on a levee, while portable pumps on tractors moved water from one part of the farm to another.[20]

In the middle of the war, the Von Glahn's planting and harvesting operation ran rather like an armored division fighting in the western desert. At planting time, sixteen Caterpillars drawing specially built seeders planted fifteen hundred acres every twenty-four hours. Crews worked eight-hour shifts, and men and machines were replenished with food and fuel in the field. At harvesttime, six harvesters drawn by Caterpillars moved parallel to one another across the lake bottom, filling grain trucks moving alongside. Von Glahn pioneered the use of aircraft in Corcoran until war regulations forced him to hand over his planes to the government. He also brought the ranch shop and machine shed to a new state of perfection. Two buildings on his ranch headquarters housed lathes, welding equipment, forges, and machine tools—enough equipment to manufacture and repair plows, harrows, specialized pumps, and other custom machinery.

In short, Von Glahn offered his contemporaries a vision of the potential of "industrial" agriculture. As a grain farmer, he was able to concentrate on the mechanization of his operations, at a time when cotton farmers around him were still struggling with farm labor problems. After the war, Von Glahn moved more in the direction of cotton, but typically he immediately began experimenting with mechanical pickers and was one of the first operators to begin using them commercially. After contributing so much to the modernization of large-scale ranching, Von Glahn suddenly gave it all up in 1948. He sold out his holdings and his machinery to a Fresno corporation and went into the oil business.[21]

Von Glahn's innovations pointed the way to new modes of farming, but new tools could not solve two other major problems that would confront growers immediately after the war and would continue to preoccupy them for many years afterwards. The farm labor problem in cotton would be partly solved by the introduction of the mechanical picker, but water control and delivery, which involved the high levels of government, would ever be a daily concern for lake bottom farmers. Both issues had a major bearing on the structure of agriculture in the Central Valley.

Technological Innovations in Cotton

The implications of the diffusion of mechanical cotton pickers in the Central Valley are difficult to assess because the data is spotty at best. From an economic point of view, the new technology was advantageous for growers. The earliest studies showed that the average savings per bale came to something in the region of eighteen dollars, compared to a bale picked by hand. For row-crop farmers the introduction of a powerful piece of machinery like a cotton picker began a process that would eventually free them from dependence on hired labor. The rapid diffusion of the cotton picker was motivated partly by a desire to solve the nagging labor problems cotton picking created. Although the introduction of pickers did not happen overnight and a certain amount of cleanup still had to be done by hand with the early machines, by 1951, 54 percent of all cotton was picked by machines.[22]

For the thousands of laborers who relied on cotton for part of their yearly income, the elimination of their jobs was something of a mixed blessing. They would no longer have to spend part of the year in dilapidated camps; on the other hand, their banishment to the outskirts of Fresno, Visalia, Hanford, and Bakersfield underlined the critical problem of underemployment in the Central Valley. In 1959 it was calculated that the average laborer worked only 144 days per year, compared to 180 ten years before. The decline exacerbated already poor living conditions for the families of farm workers. The high cost of maintaining camps, the continual criticism of conditions by the authorities, and the trouble caused by the workers themselves made most employers anxious to close down camps as rapidly as possible. As mechanical harvesters became more and more common, camps were bulldozed to the ground, and casual labor requirements were filled by labor contractors with workers who commuted from the cities.

The advent of the labor contractor came at a critical time for ranchers, when economic protection was about to be extended to farm workers in the shape of minimum wage and unemployment insurance. If their casual labor needs could be met by contractors, farmers could avoid these extra expenses. Contractors usually agreed to supply labor to a grower for a certain flat commission. The contractor made all the arrangements to secure workers and transported them to and from the work site, as well as supplying them with tools. The rancher paid the contractor a percentage of the total wage for his fee. Most contractors preferred to deal with single or unattached males because they lived in skid rows, were easy to locate, and were willing to work for the prevailing wage. Contractors took advantage of their workers in a number of ways. They were known to deduct Social Security and then not hand it in to the government. Another common practice was to charge exorbitant prices for food when workers had no other choice but to purchase a meal in the field.[23]

These developments ensured that by the end of the fifties, the prognosis for agricultural labor in cotton was fairly clear. Major structural change had seen casual labor replaced by machines, worsening underemployment and causing distress in the rural slums around the large cities. Increased use of contract labor was saving the employer outlays for Social Security payments and accommodations. Finally, the changes created a small labor aristocracy of permanent employees—irrigators, tractor drivers, and mechanics who had permanent positions and lived in rented accommodations on the ranch. Until the advent of César Chavez's United Farm Workers' movement in the late sixties, the farm labor issue lay largely dormant. It was, in the words of one writer, "a world least cared about, a world of slave markets, hiring lots, and black early mornings along skid row."[24]

Modern Water Procurement

Structural change in cotton farming and the rapid modernization of other areas of agriculture in the Central Valley had important implications for the lifeblood of the San Joaquin, its water procurement and delivery systems. Visions of vast areas of the valley brought to life by irrigation water harnessed behind dams in the foothills of the Sierra Nevada drove ranchers, agribusinessmen, politicians, and bureaucrats to push for the completion of the Central Valley Project.[25] Not only did the area lack even an elementary flood control system (in 1937-1938 and a decade later major floods had caused considerable damage) but in dry years the irrigation system failed badly too. Large-scale pumping became the preferred course of action when ditch water was not available. For the farmers of the Tulare Lake bottom, pumping was not an option on the bed of the lake, and so, many years before they had begun buying up both surface- and ground-water rights in other areas. At a time of declining resources, such actions were controversial and led to a local confrontation between the lake bottom farmers and those living near Hanford who depended on groundwater for their livelihood. This saga will be covered in a later chapter.

During the Depression came the first efforts to initiate the Central Valley Project, which would use federal resources to transport and conserve Sierra runoff in the river and canal systems of the valley below. Some of the facilities, however, were delayed for twenty years in the controversy surrounding the 160-acre limit imposed by Congress in 1902, which allowed delivery of water from federal sources to ranches that size and smaller only. Ranchers in the Kings River service area were nervous because they believed the federal government intended to enforce the limitation law, even though much of their land had been developed years before. For ten years before the opening of the Pine Flat dam in 1954, various administrations in Washington vacillated over this question. In the end the dam was constructed by the Army Corps of Engineers, rather than the Bureau of Reclamation, with partial funding from the water districts whose construction it would benefit. Great emphasis was placed on its flood-control aspects and water conservation was played down. This sidestepping of the 160-acre limit, which would have applied had the dam been constructed by the Bureau of Reclamation, frustrated critics who wanted to see the huge ranches on the lake bottom broken up. A recent interpretation of this episode has suggested that interagency rivalries played a part in the outcome. The Corps of Engineers saw a chance to demonstrate its ability to build dams and chose Pine Flat as the vehicle.[26]

The Pine Flat controversy, proved to be just the first salvo in a never ending battle that pitted a variety of groups, especially conservationists, unions, consumer advocates like Ralph Nader—the ranchers like to call professional troublemakers—against Central Valley ranchers over the 160-acre limit. Even though the Pine Flat dam was an exception in that it essentially served an area that had come into production many years before, litigation dragged on in the federal courts for three decades. A case brought by the Tulare Lake Canal Company finally ended in 1977, when the United States Supreme Court refused to hear the case. Given this lengthy litigation, it was perhaps not surprising that the large farmers of the lake bottom often appeared overly aggressive or even eccentric in their approach to water issues. At one point, when the government seemed particularly intransigent, they offered to buy the dam. As it turned out, over the years water districts were required to purchase water storage rights, so that the large farmers, in a fashion, did become part owners.[27]

In the rush to provide water delivery systems and to bring huge tracts of hitherto useless land into production, few ranchers or federal or state water officials worried much about the environmental impact of irrigation on the desert soil. By the late sixties the west side of the San Joaquin Valley was beginning to receive water through the massive California Aqueduct. As will be seen, the full implication of the transformation of the west side into industrialized ranches dependent on irrigation would only emerge fifteen years later.

The Modern Dairy

It is worthwhile to remember that in California as a whole, dairying is the economically most important agricultural enterprise, and in Kings County it is a close second to cotton. Dairy farmers in the county are the guardians of a tradition going back a hundred years, and many are related to the first Portuguese families who began dairy ranches before the turn of the century. Structural change in the industry has also benefited the Kings County dairies because urban sprawl in the Los Angeles milkshed gradually eliminated many of the large drylot dairies. Not only did Kings and Tulare counties receive much of this lost business, but they also acted as hosts to many enterprises that relocated in the Central Valley to the north.[28]

As dairying modernized in the forties and fifties, dairy farmers changed the emphasis of their production from gathered cream to gathered milk. The share of market milk between 1937 and 1941 was only 45 percent, but in the war years and after, this figure increased until, by the late fifties, fully 76 percent of all commercial milk fat production was market milk—that is, milk sold in fluid form, as opposed to milk manufactured into butter and cheese. As the drive for modernization progressed, large numbers dropped out of the business whenever regulations were tightened. In the Central Valley as many as 2,864 dairies went out of business in the fifties. An extremely competitive climate saw herd sizes increase; average annual milk production per cow went up by well over two thousand pounds from what it had been in the thirties. Traditionally, Central Valley dairies had grown their own feed and reared their own replacement heifers. As land values increased, however, farmers began buying alfalfa from custom growers and converting their own acreage to high-value crops like cotton. The raising of replacement cows continued to play an important part in many operations, but whereas in the twenties and thirties these animals would have been free to range pasture areas, now they were confined to small holding pens. Because California was out of the sphere of influence of the midwestern dairy price structure, it had its own pricing system for fluid milk, and the price of Grade B milk was determined by the price support system of federal programs. The California milk-control law required a written contract between producer and dealer that specified the quantity and quality of milk to be purchased for a given period. To retain such a contract, the dairy had to produce high-quality milk at a specified production level. On occasion ranchers might increase the size of their herds to raise income or utilize labor more satisfactorily. But if a dairy was not able to increase its quota of fluid milk, the farmer had to sell his product at a lower price on the manufactured milk market.[29]

Some California dairies employed milkers in the twenties. Even in Kings County, where employment practices had always been family oriented, commercial dairies averaged eighty-nine cows in 1959, and if family labor was not available, farmers had to employ milkers. Labor costs therefore formed a large percentage of the expenses of a dairy. In 1958 hourly wages for milkers averaged around two dollars, which hardly seems excessive. Nevertheless, this expense created an incentive for ranchers to explore labor-saving devices to cut costs. Central Valley dairies were quick to introduce pipeline milkers, which reduced milking time by 20 percent, and the smaller walk-through barn with elevated stalls and washing facilities outside, which required less labor than the old stanchion barn. But for all the innovations, the inconvenient hours and lack of time off for milkers caused very high turnover rates. In Fresno County in the late fifties, a milker's average tenure was not more than a year. The most common reason for leaving was the lack of time off the job.[30]

Of all types of farming in the Central Valley, dairying had the smoothest, most efficient method of production. All phases of the operation were made cost effective with factorylike precision. In the sixties and seventies the drive for further efficiencies would continue, and because of the economics of the business, herds would triple and quadruple, as would milk production. Average herd size in Kings was 249 in 1970, and 401 by 1982, and this county had smaller herds on average than southern California.[31]

Large-Scale Row-Crop Ranching

The large-scale row-crop ranchers of the Tulare Lake bottom provided a model for California and, indeed, the rest of the country. In the Central Valley, where the weather was of little consequence to the routine performance of operations, the land could be molded to fit the requirements of hydraulic agriculture. Tracts often conformed to section lines, and maximum use was made of the large-scale machinery available. Operations on the lake reached their zenith in 1980, when there were record profits for cotton. Then, the 1983-1984 season saw one of the worst floods in the lake's history; eighty thousand acres were under water, and two growing seasons were missed on some land.[32] By 1985 the downturn had begun to affect California. As was the case elsewhere, lake bottom farmers attempted to ride out the difficult economic times by cost cutting and searching for alternative crops to add to their rotations of cotton, wheat, barley, safflower, and alfalfa.

The late seventies and early eighties, then, provide a convenient time to explore large-scale production agriculture as practiced by one representative family partnership organized into a closely held corporation. The farm in question operated a little over eleven thousand acres, of which about a third was owned by the corporation itself and the rest was rented from family members. In 1981, at a high point for land prices in the Central Valley, most of the corporate-owned ground was valued at roughly twenty-three hundred dollars an acre. The lake bottom land alone was worth $8.25 million, not including a five-hundred-acre orange grove in the Sierra foothills. Conservative appraisals of the corporation's other assets—which included the ranch headquarters, a fifty-thousand-dollar computer, a cotton gin, a Beechcraft jet, an oceanfront home on the Central Coast, twenty thousand bales of cotton, and over a million dollars in equipment—came to almost $23 million. Because no outlays had been made for land in recent years, corporate debts were modest, with a debt-to-asset ratio of only 20 percent. According to the corporate balance sheet, net earnings averaged $221,303 a year from 1977 through 1983, with by far the best return coming in 1980, a net profit of $1,184,632.[33]

Though excellent for a number of row crops, the salty soil would not sustain many of the more exotic fruits and vegetables grown in other parts of the Central Valley. In 1981 this farm concentrated on just three crops—cotton, wheat, and safflower. In 1985, after the flood, a great deal of alfalfa was produced in an effort to return the soil to its original condition, but for the most part this farm grew a rotation of cotton on cotton, wheat on cotton, or safflower on cotton, cultivating no barley or alfalfa at all.[34]

Cotton, which was planted on 4,625 acres in 1981, was an exceedingly versatile and lucrative crop in good years. Not only did it provide fiber of high quality, it also produced oil and other products that could be manufactured into feed for livestock. The routine of cotton culture had not changed much since the forties and fifties. When cotton followed cotton in a field, the year-long schedule saw disking and ripping out of old stalks after harvest in the fall, followed by fertilizer application and repair of drains and irrigation channels. In January the ground was preirrigated, and in February herbicides were applied. Planting took place in March and April. In May the young plants were sprayed, cultivated, and hand weeded—the only time of the year when a labor force of some size was needed. In June, July, and August, the cotton was irrigated and sprayed to ward off insect damage. Finally, at the end of September the plants were defoliated by crop-dusting aircraft, so that the cotton plant leaves would not jam the cotton pickers and gin equipment. Harvesting took place in October, when the cotton was dumped from the harvesters into compactors, which fashioned large modules for transportation to the gins, where they could be stored under plastic covers until processing could take place. In 1981 this completed cycle cost between $527 and $552 an acre. The cycle for cotton planted after grain was slightly different but followed basically the same course. Wheat, on the other hand, was planted in January. Apart from weed control and fertilization in March, the only work required in the wheat fields before harvest in June was monthly irrigation. Wheat cost only about $225 an acre to produce, and this farm had almost three thousand acres in wheat. The remaining acreage was devoted to safflower. Safflower, which is manufactured into oil, makes a good rotation crop for cotton and thrives on the poorer ground of the lake bottom. After ground preparation in the winter, planting came in March. A hardy plant, safflower does not require the same kind of attention as cotton, needing only minimal irrigation and insecticide applications. It was harvested in July and August, having cost about $220 per acre to produce.[35]

Even with the production of three row crops, the routine of this type of operation was continuous. One member of the family was concerned entirely with the business side—marketing, employees, and other financial matters—while others oversaw field operations on various tracts. By the 1980s permanent employees had been whittled down to the minimum, so that only tractor drivers, irrigators, foremen, and mechanics were required. Contractors provided any hand labor that was needed, and consultants for irrigation and agronomy problems, were brought in from the outside.

Conclusion

From the perspective of the 1980s the series of "agricultural revolutions" that have transformed farming over the past fifty years, however impressive, often adversely affected farmers and especially those who worked for them, over the long haul. Since the 1930s the structure of agriculture has moved continuously in the direction of larger farms with fewer operators. This process of concentration has also seen the rise of agribusiness, which has increasingly dominated the way farming is organized. Family farms in the corn belt and the Central Valley, which had been largely self-sufficient, found it necessary to purchase nearly everything needed on the farm. Even when they were producing commodities that were sold without government subsidies, farmers became increasingly dependent on markets dominated by large-scale agribusiness; in other words, the farmer became a price taker with little hope of influencing the price at which the product was sold. New technology offered cost reductions, giving a temporary advantage to those who adopted new machines first. This treadmill effect forced farmers to adapt, to buy bigger machinery, more land, larger buildings and to obtain higher yields in the field and the milking parlor or fall by the wayside and be forced to sell out. Such a process encouraged cannibalistic concentration by the surviving farmers, while providing a market to agribusiness firms supplying the new needs of farm households.

Ill-advised government programs, encouraged by visions of unlimited overseas markets, tended to encourage planting more and more acres and milking more and more cows until surpluses in grain, cotton, and dairy products became a national embarrassment. The downturn of the eighties provided an ironic twist to these policies, for farm families were not the only ones to suffer from economic stress. The agribusiness firms that had assisted in the orchestration of change suffered equally. Thus, American agriculture found itself engulfed in a crisis that cried out for bold policies to alter the direction of farming and ranching. Unfortunately, the historical record was not promising, if land tenure policy was any guide.

3 Land Tenure

The United States has never had an effective tenure policy whose principal aim was to ensure equal access to farm operation and ownership. While the Jeffersonian legacy of encouraging farm family ownership was taken as an article of faith by the population at large and politicians and policy makers paid lip service to an open tenure system, inevitably by the twentieth century there were inequalities in the system, especially in the West, where a 160-acre farm could not sustain a family.[1] As early as the 1880s, census figures showed an apparent increase in tenancy and landlordism and a decrease in ownership that seemed to run counter to the American ideological faith that landownership in rural society should be open to all. At the time, some perceptive observers realized the closing of the frontier was partially responsible for the changes, but what they did not appreciate was that as a tenure system matured it attained a complexity all its own.[2]

Despite the centrality of the family in farming operations, no recognition was given to how the family cycle affected ownership. Instead, most observers accepted the simple descriptive theory of R.T. Ely. During the World War I era, Ely posited an "agricultural ladder," on which the landless would "climb" to farm ownership by a series of discretely segmented steps during a man's life cycle. On successive rungs of the ladder, he would work as unpaid family labor, hired hand on another farm, tenant farmer, and finally owner.[3]

Although this theory proved a useful descriptive tool for discussion of the tenure system and continued to be used until after World War II, for all practical purposes its validity was destroyed by the Depression. It was then that scholars began to grasp the full implications of the Jeffersonian blueprint for the settlement of virgin territory by family farmers. As one scholar pointed out, the corn belt (or any other territory subject to settlement by a nonaboriginal people) was really "a huge experimental plot" on which a tenure system gradually worked itself out after speculators left and the land was distributed to permanent settlers.[4]

The vacuum was filled in Iowa and in parts of the Central Valley by family farmers who were at various stages of the developmental cycle. Obviously, no two families developed identically. Some families were near maturity; others had yet to begin families; a certain proportion liquidated their farms after a few years to move elsewhere; some had many heirs; others had none at all; deaths occurred randomly. It was, therefore, a ragged, somewhat chaotic pattern that emerged as the tenure system matured. Each farm, often each parcel of land, had a complex history. In the simplest case, a family committed to farming would make a succession of transfers, each generation succeeding the next in a tidy process. But the sudden death of a family head, unsuitable or uninterested heirs, or a hundred other factors could cause a rapid turnover in ownership.

Landownership itself varied greatly from one farm to another, one family to another. In a given section, one tract might be operated by the tenant of an absentee landlord who was a widow, another by a father and son in partnership, and another by a young operator buying land with the help of a mortgage from his aged parents. A bewildering variety of chance events transformed a tenure system initially dominated by owner-operators into one with substantial tenancy and landlordism.[5]

Perhaps one of the more revealing aspects of any analysis of farm tenure in Iowa and California is the disparity in interest. Concern over what effect the structure of agriculture had on tenure in the corn belt produced a shelf of studies from World War I to the early forties, usually based on the concept of the agricultural ladder. During the same period, the California Experiment Station sponsored only two studies of farm tenure.[6] In California, it was a federal agency, the Bureau of Agricultural Economics, that sponsored studies of, among other things, the possible impact on landownership of federal water projects and the effect small and large farming had on the social structure of community (Walter Goldschmidt's celebrated *As You Sow*).[7] Thus tenure in a classic corn-belt state like Iowa was treated differently from tenure in California. In Iowa it was assumed that an open system existed, and while the state and federal governments rarely intervened to make this so, the marketplace worked in this direction until the fifties. In California the assumption was that great inequalities existed in the structure of landownership. Perhaps nothing underlines the contrast between the two states better than the publication in 1944 of a circular by California Agricultural Experiment Station giving advice to "urban business and professional people seeking to safeguard their future incomes and well-being in farming." The circular went on to discuss agriculture in the state, in the "hope of helping prospective investors."[8] In the corn belt such a publication would have been unthinkable, for there existed a deep fear of the specter of corporations bent on buying up all the land and reducing the people to hired hands with no chance of ownership. These fears still run deep.

Tenure

Before World War II, most commentators on the corn-belt land tenure system were so imbued with the notion that occupational mobility in farming was achieved through individual effort that family assistance was virtually ignored as a major component of movement towards ownership. In the first comprehensive study of the workings of the agricultural ladder little was made of the fact that the 1920 census showed 42 percent of all farm owners had bypassed two rungs of the agricultural ladder, acceding to ownership directly from positions as unpaid laborers on their parents' farms.[9] By 1920, indeed, farm occupational mobility was already influenced by family considerations. While the economy and other factors had an impact on the tenure system over the next four decades, more important was a dual system in which either the initiative for farm occupational mobility remained with the individual, free of family influence, or mobility was achieved by the family itself. In times of economic instability, as in the decade of the Great Depression, there was a shake-out in land tenure, and it became easier for an individual to begin farming without family assistance. The overhaul of the Federal Land Bank to provide low interest rates and long terms of payment also helped those without the advantage of family ties. By the end of World War II, however, the family had become increasingly important as a facilitator and stabilizer of farm careers and farm operations. In one Iowa community in the middle of one of the most progressive farm areas in the state, it was found, contrary to expectations, that familism, which emphasized family cooperation in the maintenance of property, social contacts, and occupational pursuits, proved to be the vital force in initiating and maintaining a career in agriculture.[10]

For purposes of comparison, the Iowa tenure system can be simply drawn. Although tenancy was continuously on the rise from the late nineteenth century onwards, most rented land was controlled by landlords who themselves lived either on the land or close to it. Tenants were often related to the landlord or were at least known in the neighborhood. Most farms depended entirely on family labor or, if the children were too young, on a hired man who lived with the family. By the turn of the century, land tenure in northeastern Iowa was beginning to reach maturity, something which would not take place in the Central Valley until the late Depression years. In a neighborhood where ethnic groups settled and purchased land, the process of orderly succession by younger generations began early, and tenure reflected a familial bias. However, other communities showed less stability, and families moved in and out with great regularity.

In a familial setting children worked close to home, usually without wages, until they were ready to assume responsibility on the home farm or as renters on another family's property. The key was for everyone to work for the economic well-being of the family rather than the individual. Often in the nineteenth and early twentieth centuries the more prosperous ethnic farmers aimed to buy farms for all their children, and the offspring assumed ownership of these farms at the death of the senior generation. In contrast to the closed corporate system of familism, the family that held to the individualistic ethic did not expect to contribute to any great extent to the child's career. Rather, the child bent on farming as a career left home to work as a hired man for another family, meanwhile trying to accumulate enough equipment to begin operating as a tenant.[11] It would be simplistic to suggest that no child in the familial system went to work for wages for others, and no child who aimed to begin farming by himself got assistance of some sort or another from relatives. In broad terms, however, the patterns used by these groups to achieve farm ownership were different.

The move from tenant to owner was a major step in any farm career. Again, much depended on the circumstances from which the tenant came. A familial handover was relatively simple, but someone without such contacts would have to make an extensive search for a suitable farm, as well as arrange financing. In Iowa nonrelated tenants rarely bought the land they farmed, usually because the plans of the landlord about selling and those of the tenant about buying the farm did not correspond.[12]

At first sight, tenure in the Central Valley seems entirely different, with ownership controlled by an elite and water sources so restricted as to preclude widespread settlement by families without unusual resources. In reality, the Kings River fan—where the Kings River provided natural surface irrigation when it left the foothills of the Sierra—encouraged settlement by small farmers. Especially among dairy farmers, a tenure system developed similar in all essentials to that of the Midwest. Fruit farmers required greater capitalization and at harvest-time a good deal of hourly labor. Even so, in the Kings delta much of the fruit farming was carried out on small ranches that used family resources for most of the year. Away from the area where surface water was easily obtainable, dry grain farming was practiced. The great tracts of land required were farmed by big operators on ground often owned by absentee landlords.[13]

Within forty miles of Fresno, developers began a colony system of settlement designed to bring settlers to small irrigated farms in the Central Valley. A heterogeneous collection of ethnic groups was drawn westwards to settle in these colonies around the turn of the century. Swedes, Danes, and Mennonites, many from the Midwest, were joined around World War I by Armenians fleeing the Turkish holocaust in their homeland. There were also less-structured pockets of Portuguese, Dutch, Chinese, and Japanese in the Kings River area, involved in farming as owners, tenants, or laborers.[14]

On the border of Fresno and Tulare counties between Kingsburg and the foothills of the Sierra, midwestern colonists started arriving around the turn of the century. For example, the Kleinessers, a Mennonite family from Yale, South Dakota, settled on a ranch south of Dinuba in 1910. Four generations of the family chartered several coaches on the train that transported their household goods, machinery, and livestock westwards.[15] The families who migrated from the Midwest, whether as colonists or not, were usually well established and able to buy property immediately. Other migrants came in poverty and for several decades struggled to attain a footing in agriculture. Portuguese dominance in the dairy business in Kings County is a classic illustration of how a poor group entered a labor-intensive business and exploited their family resources to the fullest extent. By sheer determination, against considerable odds, they climbed the agricultural ladder from milker to owner in a relatively short space of time.

Most of the original Portuguese settlers came from the Azores to the southern San Joaquin as shepherds. They quickly realized the potential of dairying on the irrigated east side of the valley, and even though they had no previous experience with commercial dairying, they entered the business as hired milkers and gradually learned animal husbandry and acquired the business acumen to operate successfully. Word of the establishment of a beachhead in the San Joaquin attracted a sequence of chain migrations from the Azores that continued for many years. Through time, the entire populations of tiny Azorian villages were transplanted to the Central Valley. The pattern of recruiting immigrant relatives and nonrelatives alike to work on farms was duplicated in many other places, but the longevity of this practice among the Portuguese in the Central Valley is probably unique.[16]

At this time, northeastern Iowa communities, with a more mature tenure system, experienced a transition in ownership from old-stock farmers, who had pioneered settlement, to operators who were immigrants. Around Maynard, in Fayette County, for example, the original owners began to sell out to German immigrants who either had been renting in the area or had recently arrived in this country. A transition of this kind took place in 1893, when James Means sold 198 acres to William Garnier for $7,920. The transaction was financed by a seller mortgage from Means for $6,420 at 6 percent, which Garnier agreed to pay off in six years.[17] The turnover of population and the exodus of some families with replacement by others, permitted a reinvigoration of tenure. The cycle of movement through phases of the agricultural ladder by established families was arrested, while newcomers started out afresh.

Renters and Leases

Tenancy, though more prevalent in Iowa than in California, was treated informally in the corn belt. Since there was no legal requirement to do so and they were usually of short duration, leases were rarely recorded. Indeed, only the recent farm crisis has resurrected the formal written lease between landlord and tenant. In Iowa, cash leases were used in prosperous times, for they allowed the renter a greater return for the crop and minimized the responsibilities of the landlord; in hard times, when cash was more difficult to obtain, the tenant shared some of the risk through share leases. Thus, in 1930, 45 percent of all tenants paid cash rent, but only a few years later, during the Depression, 70 percent of all Iowa leases were on a share basis.[18] When the economy improved in the forties and fifties, cash rentals returned to fashion.

In contrast, leases can be studied in great detail in California, especially before 1920, because of the way records were organized. Leases were formalized in the Central Valley because of the extent of absentee ownership, the variety of backgrounds of the tenants, the complexity and businesslike nature of agricultural production, and the advent by the 1920s of "mover farmers." These were tenant operators who specialized in growing a particular crop, and who preferred to rent rather than own land. They moved from place to place as the soil was depleted.[19]

Both absentee and resident landlords sought to rent their land to competent tenants. From the written record it is difficult to judge the extent of discrimination. In Kings County at the turn of the century Asians, Portuguese, Italians, and Anglos all signed leases as operators.[20] Unsystematic interviewing did reveal incidents of discrimination by landlords, however. Leases usually differed according to the crop to be grown. The most marked differences were between those drawn up for grain farming on the Tulare Lake bottom and those for fruit, dairy, and general farming elsewhere. The former were for a short time period, because of the risks involved with flooding, and invariably on a share basis, with a fourth or a fifth of the crop taken by the owner. In the Kings delta area fruit leases usually called for cash terms and covered periods of at least two years.

In 1900 an absentee landlord from Visalia rented Ang Haw 320 acres of prime fruit land for four years. The terms seem steep, for the lease called for a payment of six thousand dollars per year, and perhaps were an indication of discrimination. At the same time, there were no other strings attached, and if the orchard was in good condition, these terms might have been reasonable.[21] Some of the more intricate leases hedged against drought conditions. A lease between McJunken and Biddle on two hundred acres called for an increase in rent if it rained four or more inches between November 1903 and February 1904.[22] Broderick's terms with Mitssanga included not only a fifteen-hundred-dollar rental on forty acres of fruit per year but also a hundred dollars' worth of grapes and four hundred dollars' worth of apricots at harvest-time.[23]

Some of the most innovative leases were designed to help men within the Portuguese community who were just starting out. Manuel Gonsalves, for example, sold Joe Soares a two-thirds interest in all his cows and rented out his forty acres for $8,332 for a year, while retaining managerial responsibility for the operation.[24] Options to buy land and stock were another feature of leases before 1920. Anton Vierra rented 310 acres in 1912 for five years. His rental payments were inflated over the course of the lease from eighteen hundred dollars the first year to three thousand the last. In addition he was required to provide half the cream to the landlord, as well as half the calves born in the herd. As an afterthought, the lease provided the option for Vierra to buy the herd for $10,400 at the end of the lease period.[25]

Generally, cash leases were favored for fruit growing. Traditionally, harvest equipment, such as drying trays and storage boxes, were supplied by the landlord, but the majority of documents were simple affairs that stressed the method and timing of payment above all else. Grain leases for larger farmers on the lake bottom were even less specific in their terms.[26]

During the first twenty years of the century, therefore, at a time when Kings County was being opened up for settlement, renters played a key part in the development. Indeed, without the large number of immigrant farmers who often worked at some disadvantage vis-à-vis their financial arrangements with their landlords, development would have been impossible.

Travail in the Corn Belt, 1920-1940

The twenty-year period between 1920 and 1940 was exceedingly difficult for farmers in Iowa. Thousands of lives were disrupted, and the tenure system showed signs of extreme instability. These years are of

Table 2. Farm Mortgage Foreclosures in Thirty-one Southern Iowa Counties, 1916-1936

Year	Number of Foreclosures	Acreage Foreclosed	Percentage of Farmland
1915	95	11,132	.1
1916	83	11,062	.1
1917	85	12,345	.1
1918	55	6,558	.1
1919	43	6,570	.1
1920	51	7,341	.1
1921	258	42,911	.5
1922	587	98,329	1.0
1923	633	97,147	1.0
1924	724	110,069	1.2
1925	746	114,629	1.2
1926	636	97,431	1.0
1927	607	94,957	.9
1928	657	100,140	1.1
1929	621	97,914	1.0
1930	623	94,189	1.0
1931	1,033	166,531	1.8
1932	1,808	286,511	3.0
1933	1,097	170,185	1.8
1934	687	98,003	1.0
1935	650	89,775	.9
1936	582	78,900	.8
Totals	12,361	1,892,629	19.8

Source: William G. Murray, *Farm Mortgage Foreclosures in Southern Iowa, 1915-1936*, Iowa AES Research Bulletin 248 (1938), 251.

particular interest because of the current crisis in agriculture. It is often forgotten that although the full-blown Depression did not arrive until after 1930, the previous decade was also a difficult period, during which the agricultural economy was wrung out after the inflation of the late teens. Just as the inflation of the late seventies caused havoc when a deflationary trend followed in the eighties, so the farm boom during World War I fanned land speculation shortly afterwards. Many farmers injudiciously jumped on the bandwagon to invest in land at inflated prices. Unfortunately, the drastic deflation after 1920, left many of them with large mortgages and falling commodity prices.[27] In the next few years two new elements began to show themselves in the Iowa tenure system: mass foreclosures and a drastic increase in the amount of corporate-owned land.

These elements manifested themselves first in southern Iowa, where the downturn began as early as 1922, when 587 foreclosures were recorded in the lower three tiers of counties, to as many as 746 in 1925 (table 2). The increase in foreclosures occurred because those who had purchased land in the boom period could no longer pay their mortgages. The years from 1922 to 1928 were in effect a long-drawn-out liquidation of debts contracted during the years of prosperity from 1914 to 1920. In southern Iowa foreclosures never went below 550 per year, and they were spaced out over five years primarily because lenders allowed varying degrees of latitude and mortgages were not all equally burdensome. By 1931, very low prices and income levels caused difficulties for all farmers, whether they had mortgage commitments or not, and foreclosures reached a peak of 1,808 in southern Iowa in 1932.[28]

After 1934, continued difficulties in southern and western Iowa came as a result of drought and soil depletion. During this time, corporate ownership by insurance companies, banks, and loan companies became a factor in the tenure picture. In September 1933, 7.9 percent of all land in the state was corporate owned, and this percentage grew to 11.9 by 1939 as a result of bad weather and soil erosion. Although every county in Iowa contained a certain amount of corporate-owned land, less of this kind of ownership occurred in the eastern half of the state, where drought and the land boom had had less influence. On the other hand, in the southern and western counties and some northern counties over 20 percent of the land was in corporate hands.[29]

In the twenties and thirties corporate ownership was temporary and expedient. Reluctant landlords, corporations transferred the title of 615,000 acres in Iowa to private ownership between 1935 and 1937 and a further 558,000 between 1937 and 1939. At this time the contract sale was introduced. In this sort of arrangement the title of the land remained with the seller, who required a smaller down payment than was usual with a mortgage. Thus the seller retained security, and the buyer got an easier payment schedule. Insurance companies and banks also began to include lease-back provisions to former owners who were foreclosed, and later in the thirties, buy-back provisions were also utilized.[30]

Just how dramatic was the effect of the downturn of the twenties and thirties on the careers of Iowa farmers? While it is reasonably simple to use land records to trace ownership through this period, it is impossible to trace social mobility with these sources. Fortunately, a unique set of data from a Story County township spanning the interwar years and on to 1950 is available. Out of forty-four owners in 1925, only 11 percent were still farming twenty-five years later. Fully 23 percent had died as operators because they could not afford to retire; a further 18 percent lost their farms and slipped back into tenancy; and 7 percent lost their farms but again became owners. As many as 42 percent of all farms purchased between 1917 and 1925 were foreclosed, clearly indicating the difficulties farmers experienced with paying for overpriced land after the boom was over. On the other hand, none of the owners who were able to hold onto their farms until 1936 suffered foreclosure after that because very liberal Land Bank programs were in place by then. The occupational profiles of tenants were mixed. Out of sixty-three tenants farming in 1925, one-third managed to become owners, 13 percent were still tenants in 1950, 40 percent had either died or left agriculture, and 13 percent could not be located.[31]

Some areas of the state were more fortunate during the interwar years than others. In the northeastern and east-central areas, where most of the intensive work for the study took place, strong familial ties, good management and husbandry, frugal living, and careful use of credit enabled many families to struggle through. Even in the worst years, there was modest occupational mobility. In Jones County between 1932 and 1936, among 165 farmers, 9.1 percent actually saw forward impetus, while 3 percent slipped back.[32] But as in all depressions and recessions, the driving force of farm tenure, the land market, was dead.

Hard times particularly affected intrafamilial land transactions and often delayed or prevented sales by one generation to another. Landownership became less attractive for the very reason that ownership required the payment of taxes and often large mortgages. Estates that formerly might have been quickly settled were left in limbo. More use was made of tenancies in common, in which several siblings and often the widow shared ownership during a lengthy period of transitional tenure.[33] In 1935 of all farm operators in Iowa, 50 percent were tenants, 11 percent part owners, 24 percent mortgaged owners, and 15 percent mortgage-free owners. More significant, among tenants, 43 percent rented from unrelated retired farmers, widows, or estates; 23 percent from related landlords; 20 percent from corporations; and 14 percent from business and professional owners.[34]

In this context, T.W. Schultz's query "What Has Happened to the Agricultural Ladder?" at a farm tenure symposium in 1937 was intended to draw attention to the stagnation in social mobility on the farm.[35] Although tenancy appeared a rational strategy in a poor economic climate, for agriculture as a whole, a highly mobile tenantry was disastrous. It was estimated that 15 percent of all Iowa tenants moved every year in the Depression. In 1935 one-third of all tenants in the state had been on their farms for one year or less, meaning that approximately thirty-seven thousand families moved that year alone. Such turnover had ominous implications for community social and cultural development and for crop and livestock programs, as well as the wear and tear of household goods, livestock, and machinery.

Their mobility was stimulated by a combination of factors initiated by both landlords and tenants. Some landlords made unreasonable demands, assuming that their tenants could make the same kind of profits they themselves had made in the boom years. This attitude showed a grave misunderstanding of the economic and weather-related difficulties of the Depression. At the same time, many landlords were in financial difficulties themselves and could not spare the resources that would have made the farm a going concern. Keen competition for farms, with more tenants than farms to go around, also caused high turnover. Competition tended to raise rents and in turn motivated tenants to seek opportunities elsewhere. Finally, a certain proportion of tenants were poor farmers, who in whatever situation would fail and move on.[36]

In Iowa, unlike other areas of the country, there was no attempt to employ social engineering to change the tenure system. Instead, efforts were made to improve tenure by initiating longer leases, compensation to tenants for repairs, measures to fight soil erosion, to update credit mechanisms, and to raise commodity prices. It was the outbreak of war in Europe that began the process of restoring equilibrium to the tenure system.

The Central Valley between the Wars

In the Central Valley, where farm families were also hard hit in the thirties, a number of factors made patterns somewhat different from those in the Midwest. First, large farm organizations acted for farmers in the marketplace, permitting growers to act in unison rather than as individuals. Concomitantly, there was a large pool of militant farm labor organized in varying degree. The notorious Associated Farmers, which had representatives in all the Central Valley counties, was formed in direct response to actions by farm labor organizations, especially the cotton strike of 1933 centered in Kings and Tulare counties. Initially, the Associated Farmers were dominated by large industrial, commercial, and agribusiness firms such as Southern Pacific, Calpac, Pacific Gas and Electric, and the Bank of America, which had an interest in preventing labor union infiltration of agriculture. A reorganization occurred in 1936 in which farmer membership was stressed, and growers themselves were deputized to break farm strikes. In the late thirties there were several open clashes between the Associated Farmers and labor. The most violent was in Madera County, where pitched battles were fought between growers, their "goons," and striking cotton pickers in October 1939. Because the La Follette Committee published the records of all county organizations, it is possible to identify members in Kings County. These records reveal that the Associated Farmers were a fairly broad-based organization, though the membership was confined principally to old-stock ranchers.[37]

Another factor that differentiated California from Iowa was the introduction of cotton. In the midst of a worldwide depression, farmers in Kings and Tulare counties had a high-quality cash crop to market. Prosperity in urban America during the twenties also benefited Central Valley counties such as Kings, whose major products were fruit and milk. The kind of deep depression experienced in the corn belt in the twenties and thirties bypassed California. Although the state experienced a decline in land values after 1920 and those who had bought land at high prices found themselves unable to keep up with payments, foreclosures and bankruptcies were only in the region of twelve to fifteen per one thousand farms from 1926 to 1930. After 1931, land values and income declined sharply; in 1933, thirty-eight out of a thousand farms were foreclosed. Yet, for all this decline, at the bottom of the Depression, land values remained 50 percent higher than elsewhere.[38]

It was in this period that the Portuguese consolidated their position as renters and began buying up land from pioneer and absentee landlords. Their frugal life-styles, which plowed everything back into the business, depended on the labor of the whole family. Such an organization had built-in mechanisms to withstand the shocks of the downturn, and because the Portuguese relied on the family, it was never necessary for them to get involved with confrontations over farm labor.

In the fruit-producing belt of Kings County, a large work force was needed at harvesttime to pick grapes, peaches, plums, and apricots. Before the various marketing organizations were formed in the twenties, most of the labor was supplied by Chinese contractors who bought the fruit on the tree, picked it, and hauled it away to packing sheds. With the elimination of the Chinese middlemen, ranchers took over the responsibility of harvesting, either employing contractors or recruiting laborers themselves. In the early twenties Asians mainly supplied the manpower, but by the end of the decade local white labor was picking most of the fruit and then moving on to the cotton harvest.[39]

During the two decades between the wars, California agriculture attracted thousands of workers, most of whom remained low-paid day laborers all their working lives. Some, however, struggled to become owner-operators. A Dutch immigrant, James Stout, followed his brother to Hanford in 1919. For the next twenty-five years, until he bought his own dairy farm, he strove to provide himself and his family a living in a whole variety of agriculturally related jobs, illustrative of the complexity of agriculture in the area and also the flexibility and transient nature of the life of a hired man and tenant in this period.

Stout's first job was on a fruit farm working with apricots and peaches for fifty dollars per month. Stout worked on diversified farms, where he milked cows as well. He quit one job because he was required to work on Sundays. On another occasion he was employed as a picking boss for a Japanese grocery that contracted with farmers to harvest raisins. The year 1925 found him in Fresno County, employed by an Armenian raisin grower. At this job he at last saved enough to rent a place and farm on his own. The fifty-acre raisin ranch he rented for eight hundred dollars a year belonged to an old Tennessean. Unfortunately, after only one season the ranch was sold under him, and Stout had to move on. Since it was impossible for a renter to make enough to support a family in the early thirties, Stout worked for five years in the packinghouses in Hanford, where he processed dried fruit for twenty-five cents an hour. Simultaneously, he rented a dairy farm on fifty-fifty shares, and with the help of an Agricultural Adjustment loan at 3 percent, he was able to purchase some cows of his own. Despite the setback of having to destroy his herd because it failed the tuberculosis test, by the late thirties Stout was able to leave the packing plant and, with the help of his growing family, devote full time to milking. Eventually, in 1943, with five hundred dollars down and a loan from a local bank at 5 percent, he bought fifty acres from a Portuguese widow for $14,175. Thus, after almost twenty-five years of struggle, John Stout reached ownership status.[40]

Perhaps more familiar, because they were made famous by John Steinbeck, were farm laborers who migrated from Oklahoma and Arkansas. The Bennett clan, which was documented by the California Relief Administration in 1936, provides a classic example of the mobile extended family dependent on casual labor for a livelihood and on relief when work was unobtainable. Their schedule illustrates the seasonal nature of farm employment and the need for welfare assistance during the off-season. It is a situation that continues to cause problems in California agriculture to this day. In the middle thirties the Bennetts' usual yearly program was to pick cherries in Yuba and Sutter counties in May; apricots in Stanislaus in June; peaches in July through September in Sutter; and cotton in Fresno, Kings, and Tulare from the end of September to Christmas. From then until April they remained at a home base in Fresno on relief.[41]

Clay Bennett, the son of the patriarch of the clan, had worked as a young man on the farms and in the oil fields of Oklahoma. But in the middle twenties, partly because of his health and partly because of the frustrations of his life, he drifted to California. With his family in tow he moved to the Imperial Valley in 1925. There he and his sons baled alfalfa and worked on a cattle ranch. The next year they moved up to the San Joaquin and worked around Delano in Kern County picking fruit. In 1927 the Bennetts tried sharecropping cotton in Tulare, underwriting the expenses of the cotton crop and retaining a quarter of the profit. With the harvest over, they moved again, this time to Sanger, near Fresno, to a dairy farm, only to have their leased cows condemned under the new tuberculosis program. Two more years of cotton sharecropping followed. Then in 1933 low prices forced them to quit farming independently and join the ranks of farm labor. In the spring of that year they were forced to pick peas and then cotton in the fall. In pursuit of work they moved from King City to Tulare and then back to Fresno. In October 1933 Clay became involved in the cotton strike in Kings and Tulare counties. A member of the grass-roots leadership that assisted the organizers from San Francisco, he made a number of speeches that caught the notice of the authorities, and he spent five days in the Kings County jail on a vagrancy charge. (Clay Bennett was probably a model for one of the characters in Steinbeck's novel *In Dubious Battle*.)[42]

Clay's organizational abilities were put to good use in 1934, when no permanent work was available, and he served as his family's own labor contractor, scheduling them throughout the year up and down the state. The twenty-seven family members—brothers, sisters-in-law, sisters, brothers-in-law, and their children—made a sizable and pliant workforce. They traveled to their various destinations in a convoy of old jalopies and usually camped by the roadside. By 1936, when the data was gathered, the family had followed this scheme for three seasons. In the winter they all lived in a rundown twelve-room farm house fifteen miles from Fresno for twenty-five dollars per month. Even though the adult men were expert cotton pickers, they could only expect to earn $2.70 a day. All told, despite their efforts to sustain themselves, the family's total income from day labor produced a paltry $950 for 1935. It was hardly surprising, therefore, that the family was forced to spend a considerable time on relief.[43]

The seasonal nature of employment and the indifference of employers toward the welfare of workers, in season or out, created a volatile situation. In 1935-1936 only 23,000 laborers were required in the whole valley in March, but 120,000 were needed during the grape-picking season in September.[44] Workers were both anonymous and overabundant, and so growers could ignore all but their most basic obligations. As an economist from the Bureau of Agricultural Economics perceptively noted, growers "desire economic independence and freedom from government controls. They are concerned about the development of a militant farm labor class yet avoid frank consideration of the living conditions of seasonal workers upon whose labor they are dependent; they seem to wish they could farm without them."[45]

California had a low percentage of rented farms in the Depression. Overall the figure of 21.7 percent was under half that of a corn-belt state like Iowa. Even in largely agricultural counties such as Tulare, only 19.8 percent of all farmland was rented. Farms were generally larger and more valuable in California (a mean of $16,330 and 208 acres in California compared to $12,614 and 106 acres in Iowa).[46] Farming in the Central Valley required greater resources than in the corn belt; the farm occupational profile was, therefore, more selective.

While labor strife and low prices caused anxiety for San Joaquin farmers in the thirties, the cotton advantage was significant. Those who grew it were able to bring a new product in the shape of high-grade Alcala cotton to market. Not surprisingly, cotton growing became widespread on small family-run farms by 1940. The new crop apparently encouraged cash rather than share tenancy. In all 73 percent of tenants and part owners paid landlords cash. Terms averaged roughly fifteen dollars per acre, while share tenant terms called for the delivery of 20 percent of the crop to the landlord.[47] As in the Midwest, institutions—banks and insurance companies—accumulated property through foreclosures in the thirties in the Central Valley and made every effort to unload land as quickly as possible. In Tulare, only 8 percent of all cotton ground was owned by corporations in 1940. More significant was the amount of land owned by estates, absentee landlords, and widows. Roughly 44 percent of all cotton land rented by part owners and tenants was owned by absentees. Generally, absentee ownership was more prevalent on farms of 140 acres or more. As in Iowa, women were strongly represented in the landlord class. In California, community property laws presumably had something to do with boosting the numbers of women landlords. They owned 10 percent of all cotton land, although very few operated ranches themselves.[48]

In spite of the shocks of the thirties, there was a continued rationalization and evolution of the tenure system in the Central Valley between the wars. Just as in the Midwest, familism had begun to play a larger role in the way tenure was organized. There was a pronounced difference between how old-stock Americans and farmers with immigrant roots viewed the farm and their family's relationship to it. "Portuguese, Italians, and Armenian family members," one report noted, "work together on the farm, whereas the progressive Old American family members do not, but emphasize educational activities, business, and social relations."[49] In other words, although a percentage of old American families had pioneered the delta area and had managed to weather the Depression, most had only a secondary interest in remaining in farming except as landlords. In contrast, the family-oriented behavior of the immigrant farmers demonstrated a dedication to the job at hand that in postwar conditions would allow a further rationalization and consolidation, especially in the dairy industry.

The Postwar Corn Belt

Just as the farm problem of the thirties was solved by the advent of war, peace brought a new set of obstacles for policy makers to solve. Postwar difficulties revolved around production, gradually declining prices, and a surplus of farmers, who were encouraged to leave the land for other occupations. In contrast to a number of European nations, the federal government made little effort to assist those who left the land—with serious consequences in the following decades—but did continue the support programs begun in the Depression.

In the immediate postwar years observers already saw the implications of increased farm capitalization, the rise in the value of real estate, the consequent difficulties of farm acquisition by the young or others without financial resources of their own, and problems of intergenerational transfer. The structure of agriculture began to change. As land prices rose, more capital was needed to acquire title and machinery to work it. New techniques called for greater specialization, and labor-saving devices tended to lengthen the working lives of farmers, thus increasing the apprenticeship period of the young.

For the next twenty years, government policy favored the adjustment of the excess farm population from farm occupations to those in the nonfarm sector. In Iowa professionals at the Agricultural Experiment Station advocated a number of solutions to the problem of small farms, low income, and too many operators. Families were encouraged either to leave farming, to increase the size of their operation, or to seek part-time employment in off-farm jobs to supplement incomes.[50] California was on the threshold of a number of great changes involving irrigation, labor, and farm organization. The old tenure classification scheme of landlord, owner, tenant, and laborer, barely adequate before the war, no longer fit the mode of production found in the Central Valley, and over the next forty years, the face of tenure would change in the corn belt as well.

In the Central Valley after the war farming operations can usefully be divided into four categories according to how ownership, management, and labor functioned on them. In the first category were family-type farms, which were owner managed and used little hired labor. Second were tenant farms, which, obviously, were not owner managed but also did not require a great deal of hired labor. Third was a category that became increasingly important in the Central Valley after the war, the larger-than-family-type farm. These operations were owner managed but needed hired labor to function. Finally, there were the industrial-type farms, which were managed by salaried employees and used hired labor as well.[51]

But in Iowa immediately after the war, forces within the tenure system allowed it to remain in its classic form. The greater demand for farm commodities and the resulting higher farm prices in the war years enabled corporations to rapidly liquidate their farm properties. By 1945 only 2.1 percent of all farmland in the north-central states was under the control of corporations—including nonprofit institutions. Moreover, improved incomes enabled many farmers to retire their mortgages on properties acquired in the previous twenty years. At war's end, 45 percent of all farm operators were full owners.[52] Five years later 66.6 percent of all Iowa farms were debt free.[53] The latter statistic testified to the impact the Depression had made on the average operator. Debt was a scourge to be avoided at all costs. It took almost twenty-five years, and a new generation of farmers to forget this notion, which was so painfully learned.

Because of the new opportunities available in the forties, the tenure system had a resurgence. Renters could now begin saving to buy farms and, with the help of government loan programs, could purchase land that had languished in the hands of landlords and corporations. Within families, transactions were also easier. The younger generation had no fear of simultaneously making a living and paying their parents for the family farm. The earliest postwar data from the north-central states shows the effect of war and depression but also some evidence of the greater opportunities familial assistance could give. Just over a quarter of all farm owners had reached ownership by working on their parents' farms, then as hired hands, and finally as renters. Of those who had become owners without being employed in another type of farm position, more than half had spent time in the military or in other nonfarm occupations. A third of these farmers had inherited their land, while 61 percent had had to purchase from their relatives. By 1946, then, farm occupational mobility had begun to regain some of its pre-Depression characteristics, but at the same time, ascribed status in the form of assistance from the family had made an impact.[54]

Unlike some other areas of the country, the corn belt changed gradually. Indeed, a modified version of the agricultural ladder appeared in Iowa between 1946 and 1958, when 34 percent of all farm owners reported that they had progressed from being hired hands to tenants to owners, compared to 27 percent in 1946.[55] This modest resurgence of occupational mobility was presumably a reflection of the opportunities available to younger men who took the places of the elderly when they withdrew from smaller, unproductive operations. There was a modest increase during these years of those who became owners after working on their parents' farms. More important for the structure of agriculture in the state, nonoperator landlords increased from 39 percent to 47 percent, and owners who were engaged in business or the professions from 9 percent to 18 percent. Both were statistically significant increases. During this time the average size of farms grew from 141 acres to 178 acres. However, the acreage of landlord farms increased from 267 to 417 acres, or an average of 1.6 tracts to 2.3 tracts. Another change would become important as time went on: part ownership—owning some land and renting the rest—increased 35 percent between 1946 and 1958. Two different groups were participating in part ownership: young men going into partnership with their fathers and older men who expanded their acreage to take advantage of the increased capabilities of modern machinery. By 1958 family assistance in the form of gifts and inheritances made capital accumulation less of a problem for aspiring farmers. Nearly one-third of all owners obtained their land by methods involving gifts or inheritance, indicating that familial assistance has become increasingly important to the process of acquiring land.[56]

The changes that began to affect farming—new techniques and machinery and tract concentration—also influenced tenure, especially rental agreements. The increase in part ownership complicated rental and partnership arrangements. Long gone were the days when a landlord had only to obtain some assurance from tenants that they would look after the buildings and fields satisfactorily. The introduction of tiling, commercial fertilizers, insect and weed control, terracing and contouring, and sophisticated grain-storage and livestock facilities complicated the responsibilities of landlords and tenants.[57]

The complexity of both intrafamilial and extrafamilial tenure arrangements also necessitated some innovations. Business partnerships in the family were often an attempt to produce a more equitable share of the results of family labor. The conventional lease usually dealt solely in terms of gross production. A father-son partnership, however, particularly when livestock was involved, allowed a fairer distribution as far as accounting was concerned. Moreover, the granting of a partnership to members of the younger generation tended to lessen the chance of intergenerational squabbles over decision making. Another innovation was the contract sale. As income tax became a more significant consideration, the contract allowed the seller to spread his capital gains over the life of the contract.[58]

In the postwar climate of change, what were the prospects for beginning farmers and what happened to farmers in Iowa who moved to different occupations? In 1949 the careers of married farm laborers and married farm operators were compared in Cherokee County. Upward mobility among farm laborers was rare. They were less likely to have local ties in the community, and their social status was lower than that of farm operators. Although they wanted to begin farming for themselves, their prospects were poor. This material confirmed the importance of family and kinship in the occupational structure of farming. Almost all farm operators were related to landowning parents or parents-in-law, whereas the farm laborers rarely had these ties. Family help was the key to beginning a farm operation for younger renters and owners.[59]

Another study, from a statewide sample, explored the experience of beginning farmers in their quest to make a career in agriculture. Older entrants were usually single proprietors, while younger men worked in partnership with their parents. Most single proprietors were tenants or at least started out by renting land. Even if they had begun their careers as independent operators, the majority had borrowed heavily from others for many resources needed in their farming operations. For instance, only two-fifths of all single proprietors owned all their machinery, and less than two-thirds owned all their livestock.[60]

By the end of the fifties a small land base tended to hamper the progress of most entrants. Two or three hundred acres were needed to generate enough return from farming to compare favorably with competing nonfarm occupations. Most entrants farmed smaller tracts of not more than 165 acres. Those in partnership with their relatives had access to larger farms than those who were sole proprietors. Among the latter, older beginning farmers had the smallest acreages, an average of seventy-nine acres. For them the availability of off-farm work and suitable housing for their families outweighed the importance of access to larger amounts of land. Indeed, fully 64 percent of beginning operators worked off the farm during the initial year of farming. In addition, 39 percent were employed for wages on other farms, and a fifth of all wives held nonfarm jobs. In general, the smaller the farm, the more time spent in a nonfarm occupation. Not only did beginning farmers in Iowa in 1959 depend on off-farm employment to begin operations, many could not have made the move without the help of gifts and other family assistance. Although only three men had received inheritances, as many as 68 percent got some form of monetary help during the first year of farming. To underline the importance of familial assistance, about 9 percent of the total net worth of beginning farmers in their first year could be attributed to gifts. Although family assistance had always been important for beginning farmers, by 1959 even those who started out independently required some help.[61]

Movement out of farming was one of the favored solutions to the inefficiencies of postwar agriculture. As farm prices declined in the early fifties, policy makers saw the need for the rationalization of operations, the concentration of ownership, and the introduction of larger machinery to lower costs. Such changes, it was assumed, would secure a better life for all: those who left would earn higher wages off the farm, and reduced competition would increase the incomes of those who stayed.

The actual effects of consolidation were not always those predicted. In the southwestern part of the state, where most consolidations took place, it was found that forty-five of the displaced operators obtained jobs outside agriculture, twenty-five died or retired, and twenty-nine remained in farming but moved away. Ironically, of the operators who left farming, those who moved out of the state were the best capitalized, and had used the best management practices when they were in agriculture. Indeed, the farmers who left were as good as or better than the farmers who remained and consolidated. Thus, it was not necessarily the poor managers who left farming but the best educated and most highly motivated, and such people prospered in an urban environment.[62]

Most of the off-farm migrants from all over Iowa who were traced during the decade of the sixties were younger and better educated than the farm operator population. Many were tenants, already heavily committed to off-farm employment when they made the decision to leave farming. Most left to earn more money, but few had been "forced out" of farming by economic difficulties. Most could have stayed in agriculture had they wanted. Because of their firm commitment to off-farm work before leaving the land, only 3 percent experienced any unemployment. Most found jobs in the skilled blue-collar sector. Contrary to a commonly held belief, most of the migration was to small towns and villages in Iowa, rather than to distant cities and metropolitan areas. There was, therefore, a minimum of disruption of lives. Generally, most believed their move to have been beneficial. Their incomes gradually improved, and many of the women obtained employment for the first time. Indeed, it was they who expressed the greatest satisfaction with their new surroundings, but the overwhelming majority must have been pleased with off-farm life, for only 10 percent returned to farming.[63]

As will be seen, the data gathered on the exodus from farming had a special significance to occupational change in the farm crisis of the eighties. In the fifties and sixties, movement out of agriculture to suitable work in town was comparatively simple. Twenty years later, changes in the economic structure of rural America had eliminated many work opportunities. It was more difficult for farmers to make career changes in a depressed off-farm economic environment.

The Postwar Central Valley

One feature unique to the arid West was the development of thousands of acres of previously unfarmable desert. This expansion of agriculture was intimately tied to the provision of expensive irrigation water through federally funded projects. First came the completion of the Central Valley Project on the east side of the valley in the fifties, and then the building of the California Aqueduct on the west side in the sixties and seventies. The irrigation of vast acreage with public funds was intimately connected to the way land tenure was structured. Much detailed work at the Bureau of Agricultural Economics in the early forties was concerned with the potential impact of the Central Valley Project on the southern San Joaquin. In Tulare, Madera, and Kern counties 60 percent of all irrigable farms exceeded the federal acreage limit of 160 acres; some suggested that the public should have a greater say in who received interest-free water subsidies and who would farm the newly developed land. Growers and their liberal adversaries would skirmish over this question for thirty-odd years.[64]

The importance of the mechanical cotton harvester and the corresponding reduction of the farm labor force has already been noted. Although much hand labor continued to be needed in fruit, grapes, nuts, and vegetables, over the next twenty years, machines gradually took over a number of these tasks as well, especially in the tomato and nut harvests. Even more than in the Midwest the drive for efficiency entailed the enlargement of operations and the increased capitalization of businesses, further restricting entry into farming by those who did not have familial ties in the industry.

There was a movement, however, to stop the trend towards the domination of the state by big agriculture. First to heighten awareness was the United Farm Workers movement of that late sixties and early seventies. Then came efforts to make the University of California and the state more accountable in their agricultural research priorities. A committee comprising bankers, agribusinessmen, academics, politicians, representatives of private citizens' groups, and federal and state officials convened to study California agriculture. In 1977 the committee published its findings, along with some general recommendations. These urged the state to play a more active role in maintaining the competitive position of the family farm and facilitating the entry of new farmers. California should discourage the economic control of resources and markets by a few large concerns and make a family farm policy part of a general commitment to develop strong rural communities. To ensure implementation, the committee recommended that a nonprofit corporation be chartered to assist rural development and foster family farming. Probably the most important provision for family farmers was the call to provide financial assistance in the form of loans. Not since the state had sponsored two farm colonies in the twenties, which failed, had authorities expressed so much interest in family farm policy. Unfortunately, the California Small Farm Viability Project shared the fate of the earlier colonies. Apart from the founding of some corporate-sponsored cooperatives for minorities, the publication of the report proved the high point of tenure reform on behalf of the small farmer in the seventies.[65]

To understand what the hegemony of agribusiness, which was already entrenched well before World War II, meant to tenure patterns in the postwar era it is necessary to understand the kinds of farming being practiced in the Central Valley.[66] In Kings County alone, six different farming patterns are distinguishable. The first was the dairy enterprise, which by the 1960s had virtually become a Portuguese preserve but which in the seventies did receive an influx of Dutch dairy farmers, who had sold out and moved north from the urban sprawl of Los Angeles. The second type was fruit, nut, and grape farms on which cotton was grown to supplement income. Most such farms were within easy reach of Kings River water. As time went on, the small and less efficient sold out to larger operations. The third group were row-crop farms away from the Kings River. Until the fifties, many of these ranches were diversified and depended on family members to do most of the work. With the passage of time, however, they grew bigger, and competition forced the hiring of nonfamily labor. By the seventies most would be classified as larger-than-family farms. The fourth kind was the formidable lake bottom ranches, which by the eighties had been farmed for three generations. These ranches, where farmers had pioneered the utilization of the closely held corporation, could be characterized as either industrial-type farms or larger-than-family-type farms, depending on who managed them. A similar categorization applied to the ranches found on the so-called west side of the San Joaquin Valley. Finally, no tenure scheme could ignore the undifferentiated mass of unskilled and semiskilled workers, who toiled to make the Central Valley so productive.

Unfortunately, no studies comparable to those done on tenure in the corn belt after World War II were conducted in California. One of the few, which addressed the issue of increasing scale in Tulare County,

1960-1966, showed that at least half of all land purchases in the period were for farm enlargement and a further third for investment purposes as tax write-offs. Among resident operators, fully two-thirds were for operational expansion that aimed at bringing sons or other relatives into the firm. Over two-fifths of all buyers had learned that the tracts were for sale directly from the seller; about a third had bought through real estate agents. Much of the land was heavily mortgaged after the transaction, with at least a third of all sales financed through seller mortgages. Thus, land was being concentrated in fewer and fewer hands.[67]

Because of the general expansion of ranch operations, occupational mobility virtually disappeared. Farm laborers could no longer aspire to own their own farms. Only among the Portuguese was there still a degree of upward mobility. By dint of hard work a handful of Azorian immigrants could still save enough money from their employment as milkers to buy places of their own. But generally speaking, farm laborers who were ambitious—and by the end of the fifties, farm labor was becoming more and more a Hispanic preserve—left the industry to find better-paying and less onerous work off the farm.[68]

By the sixties the small rancher who owned fifty acres or less had to supplement farm income with an off-farm job, for ranches of such limited acreage were little more than hobby farms. In order to make a living in full-time row-crop farming, a rancher had to own at least four hundred acres. Many farms were far larger and often supplemented with rented ground as well. Farmers were incorporating, too.[69] Family members would assume management responsibilities in such firms, and often three generations were financially involved. This closed, family-managed organization system had its drawbacks, for in most family operations there were not enough slots on the farm to accommodate all those who wanted responsible positions. Thus, many families ended up with too many tractor drivers and not enough skilled managers. Ironically, it was the Japanese-American farm family, which had suffered the ignominy of deportation, or confinement to concentration camps in World War II, which often made the best use of its human capital. On Japanese-American ranches, only one son was normally encouraged to stay on the farm; the others obtained higher education and assumed professional positions in and outside agriculture.[70]

Nearly all the large ranches headquartered in Corcoran on the lake bottom had evolved from modest beginnings around the time of World War I or earlier to become larger-than-family-type farms. One exception was the B.J. Payne Company.[71] Whereas the Payne Company employed salaried staff in all positions of responsibility but the presidency, the remainder of the operations on the lake bottom continued to be managed by family members. It is sufficient to say that for at least six decades or more, these families became exceedingly resourceful in farming methods, use of government programs, financial acumen, and especially the acquisition and control of irrigation resources, without which their ranches would fail to produce crops. Over a long period of time their actions generated resentment not only among local small farmers but also statewide and even nationally. The source of this resentment fundamentally involved the successful circumvention of the 160-acre limitation law. For almost five decades they fought for exemptions that would permit them to make use of federally financed irrigation projects even though their farms greatly exceeded the limit of 160 acres (320 acres for a married couple). First, they ensured that a major dam they depended on for storage, was built by the Corps of Engineers rather than the Department of Interior. Second, they invested millions of dollars to buy water rights wherever they were available. And third, with some justification, they consistently maintained that the area itself was unsuitable for family farming. As they pointed out, it was impossible to live on the lake bottom itself because of the danger of flooding; moreover, the difficulties of working the land were so formidable that small farmers did not have the necessary resources to farm.[72]

Over the past forty years the closely held corporations in Corcoran have grown from organizations where the resourcefulness of the individual entrepreneur counted most to complex, departmentalized organizations requiring multifaceted expertise in agronomy, marketing, finance, and processing, as well as employee relations. As in other areas of the Central Valley, the slots for individual entrepreneurs diminished, but staff jobs were created in their place. Where forty years ago the ranch would have needed one or two foremen with a general knowledge of farm organization, the contemporary situation often requires salaried employees with specialized degrees. In the same way the permanently employed labor force is much like that of any other kind of industrial corporation; except they are not represented by a union because employers have discouraged it.

West-side ranchers had to face the same kind of criticism as the lake bottom farmers for their use of cheap, subsidized irrigation water brought to their farms by federally funded projects. From the time of World War I, corporations, as well as individual ranchers, had grown specialty crops like melons with the aid of expensive deep wells for irrigation. Compared to the east side, where the water table in 1946 might have been two hundred feet, west-side operators often had to drill their wells down to two thousand feet. Pumps capable of extracting this deep water cost thirty-five hundred dollars each. Only the largest farms could afford to drill and operate such wells.[73] As time went on, the water requirements of cotton and other crops caused the water table to drop so precipitously that by the 1960s there was great danger that the land would turn back into desert. To meet this threat a number of powerful and well-connected growers formed the Westlands Water District in the fifties to lobby officials at the highest levels of government to build a major water project with the assistance of federal and state funding.[74] Unlike the Corcoran situation, where the large ranchers had for years tried to avoid federal interference in their affairs because of the limitation law, west-side farmers acted with the knowledge not only that Federal intervention was necessary to save their farming operations but also that the importation of cheap, subsidized water might force a breakup of their properties. Interestingly, they could have opted for a scheme entirely funded by the state of California, which would not have required divestiture, but the federal scheme was considered preferable, because its water was considerably cheaper. The growers also calculated that federal water, in combination with their own wells, would provide them with the best vehicle for irrigation.

How landowners have avoided divestiture in Westlands since the water began flowing in 1968 is worthy of a major study in itself and is, in fact, the subject of several fat volumes of congressional documents.[75] As a member of Congress noted as long ago as 1955, the Bureau of Reclamation never enforced its policies very stringently. Landowners could, the congressman said, "set up corporations and partnerships, so that every adult and child in a family had 160 acres. If there are not enough of those, they could bring in uncles and aunts and, as a consequence, they spread it around, so that their pro forma title is at least within the limitation." Often this was the procedure followed by landowners in Westlands. A 1972 General Accounting Office study found that the 160-acre limitation law did not prevent large landowners and farm operators from benefiting under the subsidized irrigation program; neither did it prevent landowners and farm operators from retaining or acquiring large landholdings. There were many kinds of legal entities—such as family partnerships, trusts, and leasing—through which to circumvent the law, and only occasionally were those with the most obvious irregularities forced to make some kind of adjustment.[76]

The seventies were a decade of heightened activity in regard to the limitation law. In 1977 the Bureau of Reclamation issued a new set of rules that kept the 160-acre ownership limitation for individuals but set the total exemption for families to 640 acres. Through leasing arrangements, families could work a total of 960 acres under the new regulations. Any additional land had to be disposed of within five years, and owners were required to live within fifty miles of their property. Efforts to obtain congressional approval for these regulations failed, however.

They passed the Senate but were not acted on in the House. In the changed political climate of the Reagan administration, a Reclamation Reform Act was eventually passed in 1982. This legislation set an ownership ceiling of 960 acres for families and abolished the residency requirement. Leasing provisions remained to be negotiated. Growers were charged nine dollars an acre-foot for operating and maintenance costs on their 960 acres and sixteen dollars an acre-foot on any additional land. But the appearance of selenium in the Kesterson Wildlife Refuge in 1985, the threat of action by the Bureau of Reclamation, and in 1986 the handing down of the so-called Hammer Clause of the Reclamation Reform Act, which drastically raised water rates for farms in excess of 960 acres, indicated that in Westlands, at any rate, the struggle over land tenure was not yet over.[77]

The threat of having to pay for water at true rather than subsidized rates stimulated a further round of organizing dummy partnerships in order to avoid leasing provisions, for the Hammer Clause would specifically define what constituted a lease. The regulations, left open after the passage of the Reclamation Act of 1982, were finally published by the Interior Department in November 1986. The new regulations sharply raised water rates on excess leased land. But the power of the California water lobby—and the ranchers of Westlands, in particular—was demonstrated a few months later; after enraged growers flocked to public meetings and their attorneys met with politicians and federal officials behind closed doors, the new regulations were shelved. One rancher summed up the situation when he declared: "Sanity had prevailed. I'm relieved because it appeared for a while that the social reformers might have won over the economic realists." It seemed that once again in its eighty-odd-year history, the 160-acre limitation law had been emasculated.[78]

Conclusion

Any appraisal of the history of tenure in Iowa and the Central Valley from the turn of the century has to acknowledge several themes. Down the years landownership became an ascribed rather than attained status, with farmers increasingly coming from a smaller pool of families. Nevertheless, tenure did not evolve in a simple progression from an open system, in which anyone could attain farm ownership, to one increasingly closed to all but those who came from landowning families. In Iowa sharp economic downturns like those of the twenties and thirties created shake-downs in tenure—although these were more severe in some areas than in others. Many farm careers were destroyed, but there was also a rejuvenation of the tenure system due to the intervention of government programs and the prosperity brought by the outbreak of war. Gradually, however, by the end of the fifties structural changes in the economy increased the importance of familial connections over individual attainment for farm operators and owners in Iowa.

Tenure in the southern San Joaquin was dominated at the turn of the century by absentee owners who controlled vast holdings of undeveloped land. At the same time, in the Kings River delta, where water from gravity flow and aquifer sources permitted irrigation, small farmers, often with immigrant backgrounds, predominated until large-scale land development was initiated with the construction of federal- and state-sponsored irrigation projects. Although California had a reputation for corporate-controlled agriculture, family-run businesses were clearly in the majority in this part of the Central Valley. Corporate operations were usually closely held corporations managed by family members.

By the sixties the steady inflation in land values closed landownership and farm operatorship to all but a few aspirants in Iowa, California, or indeed anywhere else in America. As a result a new landed class appeared, mainly composed in the two study areas of third- and fourth-generation farm families whose ancestors had been European peasants. By the seventies it was virtually impossible to begin farming without significant assistance from family members already involved in agriculture. Given the increasing importance of the family as a sponsor of farm operatorship and ownership, it is important to explore how the inheritance process, which ensured the continuity of a farm business, was carried out.

4 Inheritance

Perhaps because of the legacy of the ethic of individual achievement and the obsessive concern with the agricultural ladder, few studies of inheritance were made before the 1940s.[1] By the war years, policy makers and Agricultural Experiment Station researchers in the Midwest began to realize that the tenure system was changing, and they became interested in the intricacies of intrafamilial transactions. Without academic guidance, corn-belt families had been passing on their farms to their relatives for several generations. Indeed, the classic investigation by Kenneth Parsons and Eliot Waples in Manitowoc, Wisconsin, set out to learn something from farmers who had successfully adapted Central European transfer practices to their eastern Wisconsin environment.[2]

In this dairy farming area the investigators found a willingness among parents to retire early and accept a modest scale of living. In exchange for the farm, children were required to maintain the senior generation in retirement. Services such as meals, produce from the garden, semiannual annuity payments, medical care, transportation, fuel, kitchen facilities, storage, and the use of well and rain water were often written into the legal agreement of transfer. This system of family cooperation allowed one generation to retire and the other to attain ownership. It allowed land to be discounted to a price children could afford and guaranteed payments parents could live on in retirement. It transferred the farm as a going concern and launched the children on farm careers earlier than might have been the case. In theory "bond of maintenance" agreements gave parents some security if their children failed to meet their side of the bargain. In practice the *Altenteil,* or maintenance contract, had a bad reputation by the twentieth century in Germany and eastern Europe as a whole, where the two generations were unable to abide by rigid contract stipulations. In Wisconsin, too, there was criticism of maintenance agreements that lacked flexibility. Often there were better results with transfers that were essentially real estate contracts, under which maintenance for the elderly was provided by monthly land payments from the child who had taken over the farm.[3]

The contract maintenance arrangement was common throughout the corn belt. Robert Diller's study—one of first to investigate both tenure and inheritance—found that the bulk of "disposers" of property in a neighborhood in eastern Nebraska employed the intervivos, or predeath, transfer of their real property. As many as 36 percent of all transfers between 1879 and 1937 used the predeath strategy.[4]

Some corn-belt farm families, then, had used the laws of inheritance to their advantage for many years. And while few thought in terms of a grandiose "strategy" for planning an estate settlement, their actions had allowed long-term occupation of land by succeeding generations of a single family.

In the Central Valley a number of factors made farm inheritance practices different from those of the Midwest. First, California had a community property law, which tended to simplify the inheritance process. For this reason, the kind of folk mechanisms used in the corn belt never found much practical use.[5] Second, probate tended to be handled in an urban setting because land was so often owned by absentees who lived in coastal metropolises. From the early days, therefore, there was a different flavor to California farm probate from that in Iowa. As time went on, however, there was a convergence. California's more urban and business-oriented agriculture was the ideal environment for experimentation with new forms of the farm ownership, such as the closely held corporation. Ranchers began to see a need for estate planning, and once again, California acted as an innovator on whose experience other parts of the country would modeled themselves.

In the Midwest, most experiment stations began publishing estate-planning advice in the late forties.[6] The gradual tightening of opportunities in farming after the war put greater stress on orderly procedures for transfer. Furthermore, the rise land values made estate taxes more significant. Although income and federal estate taxes were imposed as early as the Wilson administration and most states had instituted inheritance taxes by the twenties, their impact would remain minimal on most farm estates until the sixties. Before World War I, when the tenure system was still flexible, parents in Iowa were under less pressure to make decisions about the transfer of property. However, because the legal system did not automatically take care of an estate transfer, certain steps had to be taken. The farm family had to try to accommodate all its members, while making decisions about succession.

The easiest, most economical, and most efficient method in the late nineteenth century was the predeath transfer, which avoided probate altogether and allowed an owner to transfer title to a relative. Before the twenties, the intervivos transfer was illegal between spouses in certain Midwest states, but in Iowa, Wisconsin, and Minnesota, which had liberal property laws by the mid-nineteenth century, the practice was acceptable between any family members.[7] Although farmers from all ethnic backgrounds used this transfer in the late nineteenth and early twentieth centuries it was especially prevalent among farmers from Central Europe and Scandinavia, where the tradition of early retirement had developed.[8] Since there was nothing in the legal code that stipulated children must assist their parents in retirement, the senior generation employed safeguards to ensure their own welfare. The tradition of a reasonably elaborate and detailed transfer agreement was instituted along very similar lines to those of the old country.[9]

Generally one child agreed to take care of the parents in exchange for title to the land. The burden was on the child to conform to the stipulated maintenance agreement, which was often secured by a mortgage that could be foreclosed if any services were not performed. The mortgage payments functioned as annuity payments for parents. In the 1940s scholars hailed such agreements because they offered a way for family members to cooperate to keep the farm in the family and hence to stabilize the fragile tenure system.[10]

This folk approach to farm transfer had another important benefit as well. It kept the farm going as a productive operation without the disruption of probate. Certainly, disruption was inevitable if a family had to transfer property in the more conventional way through the probate court. Probate had ancient origins in English common law; it was instituted to ensure that a decedent's affairs were closed fairly and equitably, that all debts and credits were paid and all property transferred to the correct party. The drawing up of a will ensured that an estate was settled according to the wishes of the deceased, but the system could also accomplish the transfer without benefit of a will, dividing the property among the legimate heirs according to specific rules.[11]

The writing of a will constituted a commitment on the part of the farmer to estate planning, even if its object was only to transfer title to a widow. Often wills were brief documents, in which the minimum number of words achieved the desired effect of transferring all property to the widow until her death, whereupon the estate was divided among the remaining heirs. Not all wills were as simple. Generally the larger the estate, the more complex the document. But typically farmers were succinct and to the point in their instructions. Some farmers failed to make wills, perhaps because death was unexpected, because of family disagreements, or simply because of ignorance. Ironically, the estates of those who died intestate were often settled more cheaply and more quickly than those involving a will. In addition, for those with small estates and a large number of children, the division of limited amounts of property among a widow and many siblings was more easily supervised by the court than by family brokers. Until the twenties, legal transfer in Iowa used either a warranty or quitclaim deed to move ownership from one party to another. On occasion, mortgages were also used to provide income for a retired parent. In the case of a probated estate, anything from six months to a year was required for settlement. The decedent had the right to appoint an executor to ensure that the estate was correctly administered according to law and his or her wishes. Usually the executor was a son or, in the case of an early death, the spouse. If there was no will, the court appointed an administrator to carry out the same function. Although folktales of inheritance settlements would have it otherwise, the historical record in Iowa, at any rate, is unspectacular and does not reveal a trail of intrafamilial disputes.[12]

The major difference between the probate codes of California and the midwestern states was the community property provision in California. Community property was defined as all property that was created from the earnings and efforts of a husband and wife, whereas separate property was that earned by a spouse prior to marriage. Such a system had its uses in a state where by the nineteenth century landownership was in the hands of a wealthy elite, and both spouses usually brought property into the marriage. For the average operator who, struggling with his wife to create a living for the family, started out on his own, rented land, and then attained ownership, all property in the marriage would be defined as community property. If the husband died without drawing up a will, his share of the community property automatically went to the widow. Similarly, if a wife died before her husband, her share would go to the husband and, in this instance, without the need for probate administration. In estates where the husband owned separate property and died intestate, his separate property was shared between the spouse and a single child; in cases where there were several children, the widow received one-third, and two-thirds was divided among the children. Although joint tenancy was a perfectly acceptable practice under California law and was employed by many urban estates to "beat the lawyer's fee," few ranchers followed the practice.[13] Most farmers did make adequate provision for their spouses by leaving the bulk of their estate to them. Even in intestate cases, where the spouse was granted dower rights and one-third of the property, most widows were provided for. Indeed, the chief contribution of the community property provision was in eliminating most chances that a widow might be left without resources. At the same time, it gave widows access to property ownership, with important consequences for land tenure.

A number of other factors added to the complexity of California probate. Of concern were mineral rights for oil, stock in mutual water companies, and other aspects of farming and landownership peculiar to California. Moreover, the practice of shared ownership of tracts of land by several parties clouded titles. For these reasons, probate was more intricate in the Central Valley than in an average midwestern county.

Iowa Inheritance and Transfer

The system of land transfer and inheritance in Iowa before the 1940s can be traced in archival documents. In order to explore patterns in the first decades of this century, I selected a sample of 971 Iowa farmers in ten townships in 1900 from census manuscripts, probate court documents, and land records. I then traced them backwards and forwards in time, exploring their tenure, retirement, and transfer behavior.[14]

One would expect to find families in each township in every stage of the tenure and the developmental cycle. Among them, one would also expect that only a certain proportion would have remained to complete an intergenerational transfer. In the northern two-thirds of Iowa, where all but one of the selected townships lay, another, larger pattern revealed itself. In the words of an old German proverb "when the German comes in, the Yankee goes out."[15] Old-stock farmers settled the land and made improvements, but as prices rose, they began to sell out to immigrants rather than pass property on to their children. Some townships, particularly those settled early by immigrants, were more homogeneous than others. Yet because the township is an artificial, rectangular political entity without natural or community-drawn boundaries, it was rare for even the most homogeneous unit not to have "outsiders" operating farms within it.

A township with a family mature tenure system would have a core of well-established families, which would be drawn closer together by intermarriage. Some land would be operated by tenants. In a closely integrated township, these tenants would often be related to their landlords. In less homogeneous townships persistence was bound to be lower, but even in these townships core families predominated and extended their landownership through marriage and kinship. As the data in table 3 shows, predeath transfers had a certain currency not only among farmers of German extraction but also among other groups. It was usually those of immigrant origin, however, who em-

Table 3. Farm Transfers, Iowa, 1895-1945

Ethnicity	Intervivos %	Testate %	Intestate %	Liquidation %	Renter %	*N*
Yankee	15.9	18.8	8.3	17.0	40.1	277
Southern	5.0	25.0	21.6	26.7	21.7	60
German	21.8	32.2	10.1	12.2	23.4	436
British	11.1	33.4	11.2	19.4	25.0	36
Norwegian	11.3	28.8	7.5	26.3	26.3	80
Hapsburg	20.0	20.8	10.0	20.0	30.0	10
Swedish	0	0	50.0	50.0	0	2
Irish	24.3	28.5	10.0	8.6	26.8	70
All	17.9	27.2	10.2	15.8	28.7	971

Sources: US Census MS, 1900, National Archives. Probate court records and County Recorders Dockets of Deeds and Mortgages in Fayette, Benton, Hardin, Plymouth, and Van Buren counties.
Note: This table constitutes the total sample of families drawn from the census manuscripts in 1900. Thus 15.8 percent of all families liquidated their property instead of transfering it to the next generation, and 28.7 percent were renters who never owned land.

ployed it most effectively as a tool for retirement and for introducing the younger generation to farming. The classic transfer is illustrated in a deed drawn up by the Seith family, in Auburn, Fayette County:

> This conveyance of real estate and bill of sale of personal property is for a total consideration of $9,000 of which $2,000 is to be paid to the grantor as follows: By note to James Seith of even date herewith for $2,000 due $500 a year for four years at five percent per annum.
>
> The remaining consideration of $7,000 is represented by the agreed value of care and support to be furnished by grantee to grantor and for said $7,000 the grantee herein agrees to keep the grantor in his own house during the remainder of his natural life and to provide him with a good and comfortable room or rooms, properly furnished and heated for his needs, and to provide good and sufficient board, clothing, washing, and mending for said Henry Seith and to furnish all medicine and such doctors' and nurses' care as may be required by him, and upon his death to provide for and pay the expenses of his burial.[16]

This document lays out an ideal type of maintenance agreement for the aged before the advent of Social Security. Another deed, from a different township in the same county, illustrates how a similar agreement provided for the most vulnerable member of the family, the widowed mother. The following deed, dated March 4, 1887, from a Norwegian family gives an Old World solution to the problem of caring for an aged mother while passing on the farm to a son. After describing financial matters and property disposal, the deed continues:

> The grantor saves, reserves, and excepts to herself the right to use two rooms on the lower floor of the main part of the brick house on the above described premises. Also stable room and pasturage for two cows, all of which grantor is to have for and during the term of her natural life. Grantor also reserves necessary timber for firewood for her own use for the same period; some to be cut and delivered by grantee at the house, and is to have free ingress, and exgress to enjoy above reservations. The grantee hereby takes and accepts above deed subject to some reservations and agrees to provide for grantor food, and home suitable to her condition and situation.[17]

Nevertheless the most common method of passing on an estate, used by one-fourth of families, was a means of a will (table 3). Because probate proceedings went on for some time they provide the most comprehensive documentation of how a family estate was settled. The detective work is easier where there is a will, for these documents give the most explicit evidence of intentions, and periodic reports, which appear in the proceedings, add detail.

Probably the most difficult type of settlement to piece together from public documents was also the most common: the testate case in which the widow was the sole heir. Often, because ony a small number of acres were involved, the land was, for all intents and purposes, impartible. Therefore, the widow and children had to decide which child would eventually inherit the farm and how the others would be compensated. Thousands of farm families went through this procedure with varying degrees of success.

Farm wills were usually simple documents that spelled out the decedent's wishes with the minimum of fuss and bother. However, some were highly specific in content and quite complex. For instance, German farm families were often cited for their stress on perpetuating their family interests. Non-Germans, one observer states, viewed the German father "as a Tartar, working his family to the bone."[18] There was some resentment of the German determination to buy up all available land for their children; old-stock farmers, by contrast, expected their children to make their own way in the world. Some of this kind of behavior was evident in the probate materials, suggesting not only how probate worked but also the importance Germans ascribed to familism.

The Jordt family, for example, settled 240 acres of rolling Benton County prairie in October 1865. In due course John Jordt made an intervivos settlement on his wife, who completed the intergenerational transfer when her son took over the farm in 1910. John C. Jordt died in

Table 4. Types of Transfer over Time

Years	Intervivos %	Testate %	Intestate %
Before 1909	15.3	52.5	32.2
1910-19	18.4	46.1	35.5
1920-29	18.7	54.7	26.7
1930-39	13.1	64.3	22.6
After 1940	12.7	72.7	14.5
N	(174)	(264)	(99)

Source: Sample data.

1929, aged fifty-four. His real property was worth $51,197, but he had debts totaling almost $18,000 including a $10,000 mortgage. Rather than leave the chief beneficiary of the estate, the widow, to shoulder all the responsiblity of the debt, the two children of the marriage got together to find a solution. Their objectives were to keep the land in the family, to take care of their mother, and to pay off all debts. To accomplish these goals they decided to give their mother a warranty deed to her home. Her son received a deed for 160 acres of land, for which he agreed to pay twelve thousand dollars of the family debt. At the same time, the son delivered a mortage for nine thousand dollars on the property he would farm. The daughter, meanwhile, also received 160 acres, for which she paid the remainder of the debt—six thousand dollars. She, too, would deliver a mortgage to her mother for her share of the land. The yearly mortgage payments at 5 percent interest formed the annuity on which the mother would live in retirement. Thus, with some ingenuity, one family surmounted an estate-transfer crisis.[19]

Another family, the Klemmers, exemplify the German farm family tradition of working hard to accumulate enough land to provide for all the children. Charley Klemmer began his career modestly, as a farm laborer. By dint of unyielding toil and with the help of a legacy from his wife, he bought eighty acres for fourteen thousand dollars in 1904. Families like the Klemmers weathered the Depression remarkably well, and their frugality was rewarded. At his death in 1946, Charley owned 320 acres, valued at seventy-two thousand dollars, in addition to two houses in town. His strategy for the continuation of the farming operation was spelled out in his will. His two healthy sons would receive the land. In return, they would support their mother and give their sister seventy-five hundred dollars for her share of the estate. In addition, they were required to support their sick brother until he died

Table 5. Family Retention of Farms, by Type of Transfer

	⟨ 80 Acres		80-160 Acres		⟩ 160 Acres	
Type of Transfer	Retained %	Not Retained %	Retained %	Not Retained %	Retained %	Not Retained %
Intervivos	41	11	45	20	44	24
Testate, farming	49	35	28	36	33	36
Intestate, farming	5	30	13	23	7	20
Testate, retired	3	8	10	13	10	7
Intestate, retired	2	6	4	5	4	9
Data not available	0	11	1	4	1	4
N	(66)	(61)	(84)	(101)	(90)	(45)

Source: Sample data.
Note: Farms retained are those that remained in the same family for at least two generations after the transfer classified in the table. Percentage figures have been rounded.

and to provide a fifteen-hundred-dollar legacy to their nephews. Each child was given a one-third interest in the town real estate, provided their mother was allowed to live in one of the houses.[20]

Over time, as table 4 indicates, wills became much more common. By the 1940s over 72 percent of all estates in the sample were settled through wills, and only 15 percent passed through probate intestate. Will writing gradually became institutionalized. Similarly, the patriarchal model of parental behavior was largely superseded by less hierarchical arrangements, especially those of partnerships between father and son. In such cases the will played a less significant role than when it was used as the sole instrument for transferring ownership from one generation to another. However, though the actual mechanism for transfer was rationalized towards testacy, before the 1940s the intervivos transfer was the most efficient method of keeping the farm in the family over the long term. Whether the farm was small, medium-sized, or large, the intervivos transfer was the best predictor of its remaining in the family for at least two generations (table 5). Moreover, when multiple classification analysis was run using ethnicity and place of birth as control variables, the intervivos transfer was clearly the strongest predictor of continuity.[21] Using this information, I employed a logistic regression equation with a number of variables to see whether German ethnicity and the intervivos transfer of land remained significant. In this instance, age of transfer, number of acres in 1900, number of male and female heirs were entered into the equation with the dummy variables, settlement type (intervivos/other), and ethnicity (German/other). Type of settlement proved to be the best predictor,

Table 6. Logistic Regression for Predictors of Family Farm Retention

Variable	Mean	Chi-Square	P
Age at transfer	69.4	11.55	.001
Number of male children	2.5	3.14	NS
Number of female children	2.2	0.69	NS
Number of acres	148	1.17	NS
Type of transfer	.29	20.49	.001
Ethnicity	.51	12.47	.001

Classification Table

	Predicted		
	Nonretention	Retention	Total
True			
Nonretention	92	69	161
Retention	55	152	207
Total	147	221	368

Sensitivity : 73% Specificity : 57% Correct : 66%

Source: Sample data.

followed by ethnicity and age at transfer. The remainder of the variables were not statistically significant. In addition, as table 6 shows, this combination of variables proved a correct predictor of long-term ownership 66 percent of the time. In other words, an intervivos transfer by a German farmer was a reasonable predictor that a farm would stay in the family more than one generation.

While these reconstructed patterns of Iowa farm inheritance can answer many procedural and factual questions, documentary data obviously has limits that prevent a fully rounded assessment of the human side of intergenerational transfers. Because of large families and economic woes, farming became a one-generation business between 1920 and the outbreak of war in 1941. No statistic is more revealing than age of transfer for Iowa farmers. Even for intervivos transfers this was sixty-five, and for those who went through probate, it was seventy. Many sons had to wait until they were middle-aged before they had a chance to assume ownership. Furthermore, in hard times the usual stipulation that one child pay off his or her siblings for their share of the property might have discouraged some from staying home to farm. On the other hand, familism ensured that even in the worst of times some still worked to keep the farm in the family for another generation.

Inheritance in the Central Valley

In Kings County the unstable nature of tenure and the relatively late date of settlement precluded any meaningful quantitative analysis of inheritance practices. Considerable absentee ownership until the thirties both in the Kings delta and on the lake bottom negated any possibility that families with a long-term commitment to farming would be found in great numbers. Tenure needed to mature, and because the Depression further interrupted stable intergenerational patterns of transfer, it was only after World War II that any regularity could be found in the rhythm of transfers. By then, the larger operators were already experimenting with estate-planning techniques that would later be copied by farm families in other parts of the country.

Analysis of farm inheritance in Kings County centers not so much on how the process was carried out—virtually every estate went through probate, and there were no intervivos transfers—but rather where the decedent lived and whether he or she farmed full-time. In the early days, foreign estates (that is, those of decedents domiciled in other counties but owning property in Kings) dominated the probate docket. The bulk of the wealth of these estates was landed and was held for speculative purposes, but it also included such other assets as stocks and oil leases. Despite the provision of community property, all decedents drew up wills, and they invariably included heirs in the distribution who might not ordinarily have received legacies under community property law. Although none of these decedents were ranchers, their inclusion does give a feel for the speculative and vast scale of landownership by private individuals in the Central Valley before World War I.

Until 1930 foreign estates played a prominent part in probate proceedings. In 1912 the Lillis estate from San Fransisco included 3,661 acres in Kings County—all of it in the foothills of the Coast Range. However, this was only a small percentage of the total landholdings, which comprised a further 64,504 acres in other counties. Lillis left his estate jointly to his widow and only son.[22] Other estates included land in New York, the Midwest, and England. The speculative nature of most of this ownership in Kings County is suggested by the legal descriptions of the land, some of which was located in the barren Coast Range foothills, where oil would be found a few decades later.[23]

Another category of decedent had lived in the county at one time, but at death resided elsewhere. All these decedents owned productive farmland in the county and had either children or renters on the ranch when they died. When Mary Haas, the widow of a prominent pioneer landowner, died in 1910, she was living at her home on the coast. She had inherited 480 acres of some of the best land in the county, as well as half a stock in one of the mutual ditch companies, all worth $115,000. She also had 1,440 acres in Fresno County valued at $28,800, together with large amounts of cash in Hanford and San Fransisco banks. Her will called for an equal division of the estate among her children, each of whom received $49,697. The land remained in the children's hands for nine more years but was always rented out.[24] Because many families considered landownership to be purely an investment, like stocks and bonds, intergenerational continuity of ownership was uncommon. Some families did not have heirs, or the heirs were uninterested in managing a ranch, and so there was a rapid turnover in ownership.[25]

At the same time, a certain proportion of rural decedents in the county before 1920 were working farmers, and their situation was very different from that of affluent absentee owners. For example, in the first decades of the century the Portuguese were struggling to gain a foothold. A number of families went through probate when the patriarch died in prime of life, before being able to draw up a will. As was the case in Iowa, however, the court was usually flexible in arranging smooth transitions. When Jorge Rodriguez died without a will in 1900, for example, his estate included 320 acres, a herd of dairy cows, and $12,470 in the bank. This probate case was not settled until eleven years after Rodriguez's death. The court awarded his widow six-eighths of the estate and her chidren the remainder; the children, then, at the court's urging, insisted that their fifty-one-year-old mother be awarded the whole of their father's property.[26] Similarly in 1916 when Batista Franco died, his intestate estate consisted of 17.5 acres of vineyard, ten thousand gallons of wine, vineyard materials, and some horses. His children also petitioned the court to release their shares to their mother.[27] Of course, most Portuguese families did not suffer these early deaths and the difficulties of unexpected reorganization. However, it was not until the thirties that the first generation of Portuguese working farmers was ready to hand over to the next generation.

By the late twenties development in the county reached a new phase. Most wealthy probate cases were local ones. Although there were shocks in fruit farming during this period, oil finds on the west side generally boosted the incomes of the wealthy. Moreover, as will be seen, local decedents also often combined landownership with a variety of other kinds of holdings, an indication that farming was just one of their interests. Not surprisingly, the coming of the Depression took its toll on the value of estates, and land and liquid property tumbled in price. By the time the Tone brothers died in 1936, both their considerable fortunes were decimated, and all that was left was some farmland in Kings, Fresno, Orange, and Los Angeles counties. Whereas a few years previously the elder Tone would have been a millionaire, by 1936 his sons shared an estate of only $31,772. The younger brother, who died the same year, had also lost a fortune.[28]

In general the severity of the depression was reflected in both the value of estates and a predilection for intestacy, presumably because decedents were too demoralized to look after their affairs. Nevertheless, some families managed to retain continuity throughout the downturn and in some cases took advantage of circumstances. Because of the community property provision in California, women played a role in landownership they might not have had under another legal system. In 1931 Isabel Brazil, a widow, acted in the fashion of a prosperous midwestern farmer when she gave each of her children 240 acres out of the legacy that came from the community property of her husband.[29] Twenty years later, in 1954, one of her daughters died, leaving her husband substantial amounts of land in a life estate, with smaller amounts for children, and two large charitable bequests: one to orphanages in the Azores and another for masses in Azorian churches.[30]

Helen Dryden who died in 1948 also owned substantial community property with her husband, a successful businessman and rancher in Corcoran. A native of Oklahoma, he had been a small-town banker in the twenties. He then became involved in Land Bank activities in the early thirties, when that institution took over the obligations of a number of finance companies that had used land as collateral to fund the fledgling cotton business. In the late thirties Dryden had acted as point man for the Associated Farmers of Kings County, the growers' antilabor organization. He had also bought the Chevrolet dealership and had begun to accumulate some land of his own. Most of this activity took place during his marriage to Helen Dryden, and so when she died she was joint owner of the property. In all, her estate was worth $403,280, including a share of the auto agency, two thousand acres of land, and 282 shares in the family farming corporation, valued at $21,904.[31]

The Dryden incorporation, which was one of the first for a farm in the county, probably had something to do with their involvement in the auto business. At any rate, the innovation of the closely held corporation would become increasingly important from the late forties onwards and would in turn affect farm inheritance practices.

By 1947, when Shirley Benedict died intestate, the probate system for farmers in Kings County had stabilized. Benedict had bought a large amount of foreclosed property from the Bank of America in the thirties. At the time of his death he owned 4,558 acres and a total estate of $711,666. Since he died unexpectedly, without a will, and his estate was so large, there was some pressure on his widow to keep the business going. Thus after obtaining the necessary credit from her lender for the following year, she incorporated the business, and because she had no heirs, began selling off portions of the property, first to tenants, and then to other purchasers. In 1952 the estate was finally terminated, when Mrs. Benedict herself died.[32]

The history of the Benedict operation was a study in contrasts—ambitious but somewhat risky expansion during the Depression, the unexpected death of the manager without estate planning, and finally the skillful dismantling of the firm, including incorporation by the widow before her own untimely death. The administration of the estate also characterized the distance that farm and ranch probate had covered since the turn of the century. In the fifties, sixties, and seventies, farm operators would increasingly concentrate on the farm business itself, trying to perfect estate-planning techniques to ensure that the farm operation survived an intergenerational transfer.

Postwar Inheritance

The postwar years saw a gradual change in inheritance mechanisms that built to a cresendo in the late seventies and early eighties, when new tax regulations changed existing laws and made it easier for farmers to pass on real property from one generation to another without crushing penalties. The principal cause for the increases in tax liabilities on moderate-sized farms was the inflation in land prices, which began to climb steeply in the sixties and continued on this path for the next fifteen years until the deflation of the eighties. As real estate was the farmer's most valuable asset, its very illiquidity proved burdensome for family farmers trying to pay inheritance taxes. If government policy was to try to preserve the family farm, high inheritance and death taxes had the potential to make intergenerational transfer impossible. In some cases families were in no doubt forced to liquidate a portion to pay the tax bill. This unfortunate state of affairs culminated in a change in regulations in 1976, with the passage of the federal tax reform. This legislation raised standard exemptions and marital deductions enough to free many farm families from concern about estate taxes. Further estate tax benefits came their way in the Economic Tax Recovery Act of 1981, which allowed gifts of ten thousand dollars per year from both a husband and a wife to their heirs, free of gift tax. Beyond that the law allowed a $600,000 tax credit shield in gifts and bequests, to be in place by 1987.

In the fifties, when farmers were covered by Social Security, retirement became a possibility at last. Retirement has always been something of a gray area, for the simple reason that quite elderly men were capable of contributing to farm work. Landlordism was for many the only practical method of financing retirement, and therefore many farmers died owning real estate. The introduction of Social Security did not alter these patterns. Rather, the program provided a modicum of security in retirement against the worst sorts of deprivations.[33]

It would be fair to say that until the mid-sixties, there was little agitation to change the inheritance tax structure. As late as 1963, a national study found that death taxes were neither a burden to the farmer nor barrier to retention of the family farm. At that time the federal estate tax exemption was set at sixty thousand dollars.[34] In Iowa, where the average family farm was worth considerably more, the burdens of estate planning were rarely felt before 1960. For example, between 1948 and 1954 one study of a rich farming community in Grundy County, Iowa, found that only about half the total of 170 farm estates paid any kind of estate taxes. And of those subject to tax, the average payment amounted to only 1 percent of the gross estate.[35]

Nudged by the farm media, the Extension Service, and lawyers and accountants, Iowa farmers slowly started to pay more attention to estate planning after World War II. In 1946, for example, only 31 percent of all farmers in Iowa had drawn up wills, and this percentage was the highest in all north-central states. By 1958 when farmers were polled again, 58 percent had made wills, and almost all the increase was among younger men.[36]

It is debatable, of course, whether the act of drawing up a will signified a great commitment to estate planning, for most farmers little understood the complexities and pitfalls of transfer procedures. For instance, when one Iowa study asked farm couples for their transfer objectives, about one-third of the respondents were unable to provide a coherent answer. Once prompted, most cited the need for retirement income as a major criterion and objective. Equitable treatment of children in a settlement came next, and minimizing transfer costs and taxes was the third most important objective. Despite these somewhat hazy notions of procedure, 67 percent had made wills, although most had waited until they were in their middle fifties. In addition, about 20 percent owned their land in joint tenancy with their spouse—a legitimate method of avoiding probate at the death of a spouse. While most people paid lip service to the importance of retirement income, a third of all widows were given only a life interest in their husband's estates. Thus they had very limited flexibility in their later years, because they could not dispose of the property if they got into financial difficulties. Moreover, however ingenious the estate plan, some objectives would inevitably clash with others. In trying to preserve an adequate retirement income, for instance, some couples delayed giving adequate assistance to their children. Similarly, trying to provide equitable treatment to all children could run head-on into the equally laudable objective of keeping the farm in the family, especially when one of the heirs had to assume a mortgage in order to pay off others.[37]

The late forties and fifties were a period of transition in Iowa, when farmers were beginning to realize the need for professional advice in their estate planning. Before the war only exceptional families and those with a tradition of working together and passing down land from one generation to another showed much inclination to hold onto their land. In the fifteen years after 1945, these attitudes gradually became more commonplace.

By the middle sixties, farmers had to contend with rising levels of taxation. Federal estate taxes, federal transfer taxes, state inheritance tax, as well as local property taxes were all, to one degree or another, calculated from the farmer's major asset—land—whose worth continually kept rising. A USDA study calculated that the average value of production assets per farm in the United States increased about 30 percent from 1968 to 1972. Translated into dollars this increase meant that assets of a typical corn-belt hog and beef operation would have grown by approximately $240,000 per farm. During the same period death taxes increased from 1.8 percent to 10 percent.[38]

What kind of impact did inflation have on farm estates, and the taxes heirs had to pay? In 1972, somewhat early in the inflationary cycle, an Iowa study of probate calculated that on taxable estates worth between $200,000, and $250,000—a reasonable figure for a medium-sized farm—federal taxes averaged $18,193, and state tax $5,998. For a generation that had scrimped and saved through the Depression, the prospect of paying steep estate taxes was troubling.[39] The burden was made worse by the lack of liquidity in a typical operation. In 1972 Iowa farmers carried an average of only $3,274 in life insurance—in other words, 3.5 percent of a gross estate. Since the value of land and buildings in an operation was roughly 90 percent of the whole value of an estate, families who went through probate had either to sell land or to borrow money to pay estate taxes and legal fees.[40]

By the middle seventies farm interests had successfully lobbied to alter this situation. Under the new system, the estate tax was calculated as the capitalized value of the annual cash flow of the operation, or the "use valuation." This was calculated by taking the annual gross cash rental per acre, less the property tax per acre, and dividing the difference by the loan rate of the Federal Land Bank. The new method conferred dissimilar benefits on farmers with different incomes, wealth, and debt. Those with more land and higher valued land benefited more than those with smaller farms. In practice, use valuation had two main effects on landownership in the late seventies. First, the tax shelter provisions made ownership more attractive as an investment to nonfarmers, thus increasing demand for farmland and driving up prices. Second, the system tended to lengthen the time that the senior generation held onto their land, because they had to participate in its management "materially" in order to capture the tax benefits, and in consequence young farmers found their opportunities further restricted.[41]

Inheritance on Tulare Lake

Obviously the Central Valley, with its greater capitalization of operations, was especially vulnerable to both state estate and federal inheritance taxes. For this reason, more than any other, even medium-sized farms turned to incorporation in order to avoid the penalities imposed by ever-increasing land values, and families employed knowledgeable professionals to seek ways to avoid liabilities. But while farm families sought the advice of outsiders and made plans to avoid stiff tax penalties, in many instances their worst problems over inheritance would come from inside the family itself in the form of disagreement over the division of the property.

In the historical material already discussed, family disputes over settlements were rare in both Iowa and the Central Valley. Estates were modest, and paternalistic Old World traditions worked, for the most part, to promote smooth inheritance transfers. After World War II, however, especially with the boom in land prices of the sixties and seventies, together with the increased responsibilities incurred in estate planning, values seemed to change, and farmers turned to litigation to solve their inheritance problems.

Farm families have been portrayed as more disputatious over inheritance than nonfarm families with business interests. While the financial values were as great in other kinds of family businesses, the children of farmers were socialized from an early age to follow farming as an occupation. Without family assistance, it was impossible for many of them to enter their chosen career. Inheritance often represented the only chance to pursue farming. Therefore, it was to be expected that heirs who were highly motivated to become farmers would fight to secure a sufficient share of the inheritance to assure them a good start in a farming career.

All the problems associated with farm estate planning and transfer came together in prominent lake bottom families from the sixties onwards, and their patterns provide a kind of laboratory to explore the tensions apt to arise in a farm family business. Theory indicates that a family business is vulnerable to internal division because it contains two overlapping competing systems. Business and family vie for resources. Disputes and tensions can surface over hiring and territorial ambiguity in the workplace—in other words, who does what in the firm. Territorial conflict was especially prevalent in the seventies and eighties, when women began to demand a much more active role in many businesses. There are also "role carry-over" problems—that is, difficulties with fathers supervising sons or other members of the family—which could cause intense conflict and anger. And finally there was the question of control of the business and especially of succession when one generation passed from the scene.[42]

Many of these themes were encountered in the managerial and succession problems of two lake bottom families. Great stakes were involved at a time when farming was rocked first by the inflation in land prices and then by deflation and severe weather in the eighties. Disputes were serious enough to spill over into the county courts and thus permitted a closer examination of events than would ordinarily have been the case.

The dispute in which the Wood family found itself involved in 1982 was one scene in a drama played out over several decades. Despite their wealth and power, the Woods' history was "tinged with bitterness.... Tales of violence, financial daring, and affairs of the heart" dogged the family. According to a Los Angeles *Times* report, their story contained the elements of a classic soap opera, which starred a brilliant but rough-edged land baron who carved out an empire for himself only to be exiled by his two sons. The sons enhanced their wealth, but after the elder died, another feud divided the family. Charles Wood, the family patriarch was remembered in his home base of Corcoran as having "all the grace and warmth of a pirate."[43] A native of southwestern Virginia, he had come west as a newly married man with a young family just after the end of World War I. After working as a ranch foreman, he leased eight hundred acres of lake bottom and never really looked back. Over the next forty years, Charles had one goal in life: the acquisition of land. When most farmers borrowed to finance new crops, he would take the money and make a down payment on a new parcel. Not surprisingly this behavior often led to cash-flow problems, and Charles had legendary confrontations with his creditors, ending on at least one occasion with his being knocked to the ground in a bar fight. The dockets of the Kings County civil court between the late twenties and the fifties are sprinkled with cases in which the Woods are sued for payment.[44] Charles's great talent was to bring marginal lake bottom land into production and to ensure that this land was supplied with water through an elaborate dike and canal system. The family even obtained the part ownership of water rights of the Pine Flat reservoir in the foothills of the Sierra. "Cockeye" Wood—he had a glass eye—eschewed the trappings of wealth; dressed in a slouch hat and old coat, he drove around his ranches in an old Cadillac with his collie on the front seat. According to his daughter-in-law, he was a frustrated actor. He slept little, commonly worked ninety-hour weeks, and carried around most of his business transactions in his head, committing those of greatest importance to pieces of paper stuffed into his back pocket. He was a prodigious drinker and kept up a liaison with a Hanford women for years, apparently with the knowledge of his wife.[45]

Despite his eccentric and often violent behavior outside the family, his bitterest dispute was with his two sons, Leonard and Tom. During the final twenty years of his life, they battled over the management of the family corporation, and their disagreements still divide the family. Basically, the conflict revolved around decisions concerning the future of the business. Losses from cattle, and the control of the all-important cotton allotments (his sons were terrified that he would remain in control of the quotas, which were vital to the business) were just two of the items that divided the generations. The dispute was serious enough for Crocker Bank, the family's lender, to threaten to cut off funding if Charles was not removed from the board of directors of the closely held corporation. In July 1961 the whole Wood family signed a document with the intriguing title of "Wood Family Compact—Interim (Operation Armistice)" in an attempt to resolve their differences and provide a mechanism for operating the Wood companies satisfactorily.[46]

In memoranda filed for civil cases, which surfaced after Tom's death in 1982, it is possible to trace the evolution of the intricate family estate plan devised to provide for "prudent" inheritance and gift and estate tax planning and to facilitate the distribution of family assets to the next generation, as well as to ensure that there was a continuity in the operation, that stock would remain in the family, and that liquid assets were available to pay inheritance and estate taxes.[47] The first estate plan was drawn up in 1958 to cover the younger generation. In the following years concern for the brothers' health saw the creation of corresponding and complementary provisions for both their families. Tom was named executor and trustee of the testamentary trusts created by Leonard, and vice versa. Throughout the early sixties, these provisions were continually updated. In 1963, for instance, a voting trust agreement recognized that the management services of Tom and Leonard "were essential to the continued success of the company" and that bank financing was dependent on their continued management. By the late sixties a number of life insurance trust agreements were drawn up to ensure continuity in ownership. Most important, the agreements were arranged so that the survivors could purchase the stock of the deceased and reimburse family members for any losses the death might have incurred.[48]

From 1966 to 1972 there were protracted negotiations directed at a possible permanent solution to the intergenerational conflict between the Wood sons and their father. These proposals included the possible sale of Leonard's interests to his sons, and a split off of other assets to Charles. These matters were finally resolved in 1972 by a written agreement under which Charles and his wife sold their stock to the company, a $3.5 million debt was forgiven, and an island in the Sacramento delta was sold to Charles. He had actually been exiled there in 1964, and this was a formal recognition of that fact.[49]

It would seem that hardly had the Woods solved one intergenerational dilemma when another presented itself. By the middle seventies a third generation of the family was beginning to take an interest in the business. This was reflected by the creation of a series of complementary trusts and interlocking agreements to streamline their common estate plan and to preserve continuity in the management and control of the Wood companies. These interlocking trusts included the third generation of the family as beneficiaries and cotrustees of their parents' estates. As before, each trust was organized so that the net proceeds on the life insurance policies of the decedent would purchase the stock of the company held by the estate of the decedent, and each of the four Wood parents had identical trusts. For example, Leonard Wood's two daughters were beneficiaries of his trust; Tom Wood and Leonard's daughter Jane were cotrustees; and Leonard's wife, Evelyn, was the recipient of a life insurance policy designed to cover the purchase of company shares. Although at various times individuals from outside the family held shares in the company as compensation for services rendered, the basic premise behind all these well-laid plans was retention of family control of the company. Indeed, the trust agreements provided that the surviving brother would run the company until his death, and then control would pass jointly to Tom's eldest son and Leonard's eldest daughter.[50]

When Leonard died of a heart attack in 1982, however, it quickly became obvious that these carefully nurtured plans would not be followed by his immediate family. They refused to sign the necessary documents to set the previously agreed plan in motion. Tom summarized the situation when he wrote to his sister-in-law a month or two after his brother's death:

> The problem that I now face is that you have stated your intent to deviate from our plan, and hold your stocks separate from the testamentary Trust contained in Leonard's will. The effect of such a decision is that if I predecease

you, you become the controlling shareholder until your death. Furthermore, if Ann [Tom's wife] predeceases you, and I am alive, you could conceivably assume voting control. In addition, after either Ann's or my deaths or both, you could change your existing will, and my family could be cut out in perpetuity. This is not what either Leonard or myself intended, nor does such a result conform to our family's agreements.

Evelyn, I am extremely concerned about this, so much so that I feel compelled to address you in writing. I am especially concerned because I am very worried over Janey's [his niece] attitude and intentions. Some weeks ago Janey demanded that she be made a vice-president in charge of administration immediately and be paid a salary of $54,000 net. About a week ago she demanded such a position again and said she had to be paid a salary of $75,000. At that time she threatened that if I didn't comply, she would give the story to the L.A. *Times*....

Evelyn, this is nothing more than a form of blackmail. I don't intend to make a decision based on such threats nor be pushed by such tactics, and despite these actions, I intend to give Janey every chance in the company, pay her at a rate commensurate with her position and ability, and move her along as quickly as circumstances merit. Leonard started her at a salary of $25,000, incidentally unknown to me, and advised me she would be working at the computer department. I must tell you, however, that in the months proceeding Leonard's death, he was very upset with Janey, and expressed to me serious reservations about her judgment and capabilities. Based on Leonard's observations, and my own of the last month, Janey in terms of ability, experience, and training, is simply not ready to assume the responsibilities she demands....

The present situation must be addressed quickly. Life is uncertain and my objective is to confirm our existing plans, and put them in a totally enforceable position as soon as possible.[51]

Tom terminated his appeal by apologizing for his handwritten letter, pointing out that in view of its contents, the fewer people who read it the better. Actually it did not take long for word to spread to the press statewide that the Woods were fighting among themselves again.[52] It would seem that Leonard's widow was serious in her desire to take an active role in the management of the company. However, her brother-in-law and nephew were able to outmaneuver her at a board of director meeting four months after her husband's death, effectively taking control of the company. Over the next two years in suits and countersuits the family battled over the executorship of Leonard's estate and the control of the company. In the words of one lawyer, there "was total hostility and adversity between the two branches of the family." The preamble of one suit accused Evelyn and her daughters of deliberately, maliciously, and spitefully committing acts in disregard of the rights of the other branch of the family and asked $5 million in damages. As is often the case in such a situation, the court denied a number of requests, thus defusing the issue and turning both parties towards a settlement.[53]

On the face of it, the Wood fight could not have come at a worse time, for the spring and summer of 1983 saw one of the most serious floods in Tulare Lake history. While the family was spending thousands of dollars on lawyers, a large portion of their property was under water, and no crops could be planted. Although they did not know it, the farm debt and commodity price crisis were also beginning to affect the Wood operation. The company had bought thousands of acres of high-priced land a year or so earlier, adding to the uncertainty caused by the changing of the guard in management. Drastic cutbacks in the labor force as a solution to economic problems backfired and triggered the first appearance ever of representatives from the United Farm Workers in Corcoran. The Wood Company had to allow an Agricultural Labor Relations Board-sponsored election to decide whether its workers would be represented by the union.[54]

Given this situation and widespread rumors that the company was in financial difficulties, it was not surprising that the family fight was quietly settled in the spring of 1985. Leonard's side of the family settled for a cash buy-out for something in the region of $6.5 million. In view of the deteriorating economy, this was a shrewd strategy. Evelyn Wood received about $1.5 million, and each daughter a slightly smaller sum. A small amount of property also changed hands, as well as lifelong medical insurance coverage for two of the heirs and, for the widow, a membership to the Monterey Country Club.[55]

In their actions, which were often larger than life, the Woods obviously were not typical of the average Central Valley family. At the same time, elements of their elaborate estate plan and their intrafamilial clashes illustrate the difficulties involved in keeping an operation going and the lengths to which family members would go to obtain what they believed to be theirs. The second example of an inheritance dispute was perhaps more representative of ordinary operations, in that it initially involved the inability of brothers and sisters to agree on a family partnership termination—a common enough occurrance in modern agriculture.

Unlike the Woods, the Shepards had an unblemished reputation in Corcoran. The famly difficulties originated in the failure of the five branches of the family to agree on the terms of the dissolution and separation of a busienss that had been in operation for over twenty years. Shepard Farms was founded by Jim Shepard in the twenties, but it was one of his sons, Matt, and his wife, Louise, who successfully expanded the operation after World War II. In the fifties Matt's two sons and three sons-in-law joined him in the extensive farming operation around Corcoran and in Tulare County.

Because Matt Shepard suffered from Parkinson's disease for the last fifteen years of his life, his elder son, James took charge of financial matters, while his other son, Steve and his son-in-law Joe Quinn ran the field operations at Corcoran; Andy Slater, his eldest daughter's husband, operated the orange ranch in Tulare County. After Quinn died, in 1979, his widow came into conflict with her brothers because she believed she was not receiving a fair rate of return for her share in the operation, which constituted her sole source of income. Negotiations virtually broke down in 1981, when James and Steve Shepard and Andy Slater took the position that they would be the final arbiters of how the property would be divided. When Deborah Parrish, the other sister, supported Carol Quinn, the battle lines were fully drawn.[56]

As is often the case in sibling farm partnershps, the tensions surrounding the employment of the younger generation exacerbated the troubles. Carol Quinn's son, who had worked with his father before he died, encountered what were called "problems in effective communication" with his uncles. David Quinn, described by one of his uncles as a "whippersnapper" and a "pipsqueak," did not like to take orders from his relatives and was highly critical of their management. "We couldn't do anything, regardless of what it was," said another member of the family; "everything was wrong: the irrigation system, farming methods, everything." These disagreements eventually caused David's termination, and afterwards, according to the other family faction, he became the "instigator," who motivated his mother and his aunt to "seek more than their entitlement" from the division of the operation. One further actor in the drama, was the family matriarch, Louise Shepard, who not surprisingly became distraught when she discovered that the farming business she and her husband had taken years to establish was about to be the subject of family litigation. Her two daughters had decided to take their brothers to court. Louise Shepard remained a key player because she had inherited from her husband some land separate from the property of the farm corporation, which she could use as a bargaining chip in the family power struggle.[57]

In the summer of 1982, Louise Shepard went to the family attorneys in Fresno to alter her will, reflecting in the changes her attitude to the dispute. No longer would there be an equal division of her land between children and grandchildren at her death. Instead, she disinherited the daughters and the grandson who had questioned the legitimacy of his uncle's control. Hardly a year later, the old lady died, leaving a disputed will and an estate that, including the land, was worth roughly $4.5 million. Not surprisingly, the daughters filed suit in civil court. They disputed the legality of their mother's final version of her will, citing the influence brought to bear on their mother not only to coerce them to drop their original litigation but, more seriously, to change her will. The daughters accused their brothers of transporting their mother to Fresno and holding her hand while the lawyers made the necessary alterations. This disinheritance suit, which centered on four hundred acres of prime farmland, worth over a million dollars, was eventually settled out of court, but not before the whole family had spent several days at hearings open to the public and a number of lawyers had argued the merits of their clients' cases before the local press. It was the state of the farm economy, together with the fact that much of their property was under water in 1983-1984, that forced the family to take a realistic look at their circumstances and to urge their legal counsel to achieve a compromise. Thus by the end of 1984 a rough value had been assigned to the various properties under dispute, and the process of dividing up the ranches into separate entities was tackled. As the Woods had discovered, 1984 was not a good year for a farming corporation to be engaged in a family dispute.[58]

These two cases were played out against a background of soaring, then plunging land values. The agricultural provisions of the Economic Recovery Act of 1981 were designed to further alleviate the liquidity problem of farm estates by increasing exemptions to the federal estate tax. Unfortunately for agriculture as a whole, the legislation would prove of little value in the long term, for among other things, its income tax provisions helped bring about the cycle of deflation that would prove so disastrous to farming in the middle eighties. When families that had built a strategy of expansion based on land equity were left holding heavily mortgaged land in a declining market, questions about inheritance strategies became academic. They were concerned about holding onto their property in the very short term and about the chances of extending their loans for another year.

Conclusion

Historically the demand for family continuity on the farm in Iowa was a cultural characteristic brought to the Midwest by European immigrants who were determined to remain in farming regardless of the economic forces and structural change going on around them. In the Central Valley a rationalized farm inheritance process after World War II grew out of a desire to preserve the farm as a business entity. Hence, California farmers emphasized estate planning and the formation of family-controlled corporations while agriculture in California was expanding, and larger firms needed protection from inheritance taxes and exposure to liabilities.

In the Midwest, agricultrual economists had looked to Old World patterns of land transfer in the hope that they would learn something about reinvigorating a tenure system made moribund by the Depression. What seemed valuable to researchers was the ability of the family to work together on the farm and eventually to achieve a smooth transition to the succeeding generation, but clearly such a system relied on paternalism and demanded considerable sacrifice from heirs who had to wait to take over the farm.

Increasingly after World War II, urged on by extension agents and the farm press, farmers and ranchers grew more accustomed to thinking of their farms not as one-generation operations but as multigenerational firms requiring intricate planning in management and finance. Many operations became family businesses, in which family members had to work together in an environment that lacked the traditional checks and balances that had formerly preserved continuity. Thus, new strategies for estate planning and inheritance became part of modernization in farming. At the same time, human relations, not agronomy, often became the Achilles' heel of a farm family business. In addition, the multigenerational farm operation was vulnerable in an economic downturn, especially if it had borrowed heavily to expand in order to accommodate heirs. Unfortunately, the use of credit, like the weather, was a necessary risk farmers routinely had to bear.

5 Credit

Without credit farmers cannot function. Cash flow is limited for all but certain specialists such as dairy farmers. Even in the late nineteenth and early twentieth centuries, when the egg money often provided farm families with a steady source of income, only the most frugal farmers avoided borrowing money for land purchases or operating and living expenses. Credit influenced land tenure and inheritance and, like them, was affected by the changes in the structure of agriculture in the twentieth century.

Much borrowing was local and noninstitutional—fathers and sons, between neighbors, or through seller mortgages. Farm families, are, to use an economist's term, heterogeneous in both family structure and financial needs. Some are relatively well off; others are hard up. Often borrowing and saving behavior depends on what stage in the farm cycle a household has reached. Younger farmers have to borrow heavily, whereas older operators have savings to invest. Some types of farming have different cash-flow projections than others. Moreover, some family members might work off the farm, thus boosting income and reducing the need for credit.

Obviously the financial affairs of any family are a matter of some delicacy. In times of economic stress, when lenders and borrowers are often forced into tight corners, emotions get the better of reason. Borrowers feel that lenders have taken advantage of them, and lenders are subject to abuse and name calling by unhappy borrowers. It is all to often forgotten that lending is a two-way street. Farm mortgages not only satisfy the borrower's need for funds but also the creditor's desire for a reasonable investment with as little risk as possible. Institutional borrowing at the local level is a face-to-face proposition. After all, the whole purpose of the local institution is to provide opportunities for investment by neighborhood people, whose money is then loaned out to farmers. The small community has always depended on the trust and cooperation of its citizens for prosperity. Quite often local bankers and abstract company personnel acted as agents for large commercial loan and insurance companies, so that even relatively anonymous credit sources had local agents who needed to satisfy all parties to succeed in their business.[1]

In late nineteenth-century Iowa credit mechanisms were especially intricate and locally oriented. Unfortunately, few records of short-term loans, either for consumption or production, have survived, and the ledgers in which chattel mortgages were recorded have largely been destroyed. Only in the 1930s does secondary literature provide details of lending practices for non–real estate loans.

In the Central Valley, short-term production loans and chattel mortgages, faithfully recorded in the courthouse, provide a record from the turn of the century onwards. Of particular interest are crop mortgages, which came to the Central Valley when cotton growing was first introduced in the twenties. These loans were specifically designed by cotton producers to allow growers to plant a crop, cultivate it, pick it, and then sell it to the producer's gin. From the middle twenties, when the first gins were opened, until the fifties, when other sources of money came available, crop mortgages were the principal source of credit for cotton growers. In the Central Valley, a chattel mortgage often accompanied the crop mortgage and helped to secure the loan. On cotton farms, land and machinery acted as security; on a dairy, livestock and land usually served this function.

It goes without saying that the borrowing of quite large sums of money incurred considerable risks for a farm family. Personal problems, the state of markets, and the weather—all had a bearing on whether a particular loan could be paid off on schedule. Some families shied away from these pressures and made it a rule never to make a purchase unless they could pay cash for it. Others objected not so much to the risks involved as to the interest payments, which ballooned the price of a piece of land or machinery. Over the years, however, these conservative farmers were apparently in the minority or were bought out by risk takers, who periodically had to pay dearly for bouts of overindulgence. In Iowa, particularly, memories were short. The Great Depression was preceded by a period of rapid inflation, followed by severe deflation in land values that formed a backdrop to a period of extraordinary difficulty for farmers.[2] The Depression taught farmers a number of painful lessons about deficit financing. Yet fifty years later the scene was repeated.

Given the salience of equity financing in the current farm crisis, it is worth beginning by exploring in detail the previous occasion when family farms were taken through the credit wringer. Before we turn to grass-roots borrowing practices in both localities, a unique set of data from central Iowa permits a long-term overview of the place the real estate mortgage played in one county's agricultural development, especially during the twenties and early thirties.

Mortgages in Story County, Iowa

In 1933 William Murray and his collaborators studied real estate mortgages in Story County, Iowa, to discover the results of short-term lending cycles. During the settlement period, from roughly 1850 to 1880, they learned, mortgage debt had mounted rapidly. Settlers needed credit to purchase land, to make improvements, to buy livestock, and to pay operating expenses until they were able to stand on their own feet. Not surprisingly, the number of mortgages rose dramatically. By contrast, the period 1880 to 1910 saw a steady increase in land prices and only a gradual increase in outstanding debt.[3]

Land mortgages were primarily used in three ways: to buy land, as security in a transaction direct from a seller; and for renewal. Land purchase cycles coincided with peaks in mortgage transactions, as in 1875, 1891, and 1920. Renewals, on the other hand fluctuated in almost inverse proportion to land purchases, because they occurred during hard times, when mortgages came due and had to be renewed because income was not sufficient for payment. Similarly, foreclosures came after several years of heavy mortgage activity, and their greatest incidence occurred in the years 1876-1879, 1923-1928, and 1931-1932.[4]

The Story County data also provide information on lenders and where they came from, as well as interest rates and terms. Mortgages were supplied by several sources, including private investors, insurance companies, deposit banks, former owners, land banks, the school fund (which provided financing for schools from the sale of land), and brokers. During the frontier period, the schoolfund rapidly dried up, and private investors supplied as much as 75 percent of all funds. Even by 1930 private sources still represented 18 percent. Former owners remained a force throughout the seventy-eight-year period, while commercial lenders, such as insurance companies and banks, gradually intruded on the territory vacated by private investors. The actual value of debt, however, tells a somewhat different story. During the nineties, for example, total debt in the county expanded $1.1 million. Private investors increased their holdings by $481,000, as compared to $335,000 by insurance companies. Between 1900 and 1910 the debt ran almost $3 million. In these years insurance companies became especially active, increasing their share by $1,651,000. Deposit banks held $649,000 worth of mortgages, followed by private investors and former owners, with $309,000. Between 1910 and 1920 the patterns altered again and a further $17 million in debt was added. Former owner loans rose to $5,341,000, with almost half of it in 1920 alone. Private investors registered a total of $4,646,000, with insurance companies carrying $3.8 million and banks $1.3 million each. In the next decade another reversal took place when insurance companies and banks took over the losses of private investors and carried out considerable refinancing.[5]

Although the state legislature pegged official interest rates from a high of 10 percent in the 1850s to 8 percent in 1890, unofficial rates were certainly much higher. Nevertheless, by the turn of the century rates had stabilized at around 6 percent. By the start of the Depression they were even lower, usually not more than 5 percent. Terms for loans in the nineteenth century were normally five years. Installment-type loans were favored in the early years, followed by balloon loans, which had to be paid in one lump sum when they were due. By 1910 the advent of high land prices had made long-term installment loans of at least ten year's duration more common. The Land Bank also introduced thirty-year loans just before the outbreak of World War I. But if borrowers welcomed very long terms of repayment, in general lenders still preferred the short-term arrangement, because funds remained more liquid.[6]

Boom and Bust in the Twenties

These data not only provide a sound introduction to overall lending and borrowing trends over a long period, they also focus on a issue that has returned in a dramatic way to haunt farm families in the 1980s—the speculative psychology that accompanies a boom and its impact on the farm economy if a bust follows. I touched on the effect of the most dramatic boom and bust previously experienced in Iowa in the discussion of land tenure. After the 1919 boom there was a lag in time before foreclosures became common in southern Iowa. Although prices dropped in 1921, foreclosures did not increase materially. Many considered the drop temporary, and they looked for a return to the prosperity of a few years before. In the words of William Murray, "As the years 1922, 1923, and 1924 passed without a return to the previous high levels, owners with heavy mortgages and lenders on these same mortgages concluded there was no escape from debt liquidation. The ten-year period of continuous foreclosure, therefore, was in effect a long-drawn-out liquidation of debts contracted in the previous prosperity period. The spacing out of the foreclosures was in part the result of the varying degrees of optimism held by different lenders and owners of heavily mortgaged land, and in part a reflection of the differences in mortgage burden—the mortgages for larger amounts per acre being foreclosed in the earlier years."[7] In the 1980s, such lag patterns began to reappear, although they were tempered by government intervention that slowed the process.

Although there was little connection between the boom foreclosures of the twenties and the depression foreclosures of the early thirties, there was at the time a widespread belief that avarice had caused the whole problem of farm depression in the early thirties. Many believed farmers were getting only what they deserved for having speculated in land during the wartime period of high prices.[3] It is true that farmers themselves were the most frequent buyers of land during the boom. They constituted 65 percent of all buyers, while real estate agents, bankers, and commission men represented less than 10 percent each. Nevertheless, analysis of the boom showed that they had not been purchasing irresponsibly. During the period, farm income was up as compared to land values for expenses had not increased as rapidly as the selling prices of farm produce. Farmers, therefore, had more money to invest, and because banks had a greater volume of deposits, credit was easy to obtain. There is no indication that standards of security were lowered, only that funds for mortgages and personal credit, especially those provided by sellers, were abundant. Many of these sellers were retired farmers, whose retirement had been delayed by the war, and who preferred to convert their equities into seller mortgages, rather than have the responsibility and risk of renting their land.[9]

Once the boom got under way at the war's end, and it was clear that land prices were rapidly rising, many who otherwise would not have been interested in the market bought land as a short-term investment. From one-fourth to one-third, of all pruchases were speculative. Some farmers had delayed buying for any number of reasons. Now, fearing that prices would rise too high, they jumped into the market. Others shrewdly held onto land just long enough to take their speculative profits.[10] This was the time when real estate dealers rode about the countryside stirring up excitement, and farmers were so busy "sitting around the livery stable and playing games with options" that no time remained for field work.[11] The ultimate consequences of the boom were of course very similar to those sixty years later. Even if deflation had not occurred, the relationshp of farm values to farm incomes was such that it was doubtful whether many farmers could have paid for their farms with the kind of return they were getting for their produce. As it was the sharp deflationary cycle began in the fall of 1920, and the farm economy started on a trend from which it would not recover until the 1940s. Farmers, moreover, were not the only casualties; the twenties were a terrible time for agricultural banks in Iowa as well. From

1921 to 1929, no less than 528 banks failed in the state, 302 in the worst years 1924, 1925, and 1926. The delay in the failure of these institutions was directly attributable to the downturn of the agricultural economy and the long-drawn-out liquidation of debt after 1920.[12]

Grass-Roots Credit in Iowa

In the majority of Iowa townships in my sample, commercial credit remained relatively unimportant until well after the turn of the century. Initially, homesteading and railroad purchases allowed families access to land with a minimum of expense. If a mortgage was needed to purchase land in eastern Iowa, funds were often obtained through a so-called eastern mortgage. Private investors in New England, New York, and Pennsylvania used the services of an agent in the Midwest to invest in corn-belt farms. As communities gradually matured, a close-knit web of familial and neighborhood borrowing came into existence to supply the young with the credit they needed to buy land and older established farmers with secure investments. Sometimes, local landowners and other notables would club together to found a bank to serve the community. Such was the case with the Zeigler family in West Union, Fayette County, who began an extensive mortgage and loan business in the 1880s before turning to commercial banking.[13]

In the Catholic community of Auburn, the local priest took an interest not only in his parishioner's spiritual lives but also in their economic well-being. In the last two decades of the nineteenth century he made a point of counseling them on real estate opportunities and even acted as an intermediary in a land sale.[14]

The role of the broker, who matched lenders and borrowers or sellers and buyers, was important in a relatively unsophisticated economy in which lending was not yet institutionalized. After the turn of the century the local barber performed this function in Auburn. For many years he would not only give local farmers their haircuts but at the same time provide short-term consumer loans at high interest rates. In ethnic communities, where land trading and the purchase of small tracts was constantly occurring, short-term loans often were not officially recorded. Like intermarriage and kinship, local lending arrangements between neighbors and relatives served to reinforce community solidarity.

With the use of some early data from Wisconsin, whose economic climate resembled Iowa's it is possible to look at credit transactions between merchants and farmers before World War I. In Wisconsin dairy country, store credit was universal. Merchants carried "as much on the books as in stock," and a study of 110 general, machinery-

Table 7. Source of First Land Acquired, 1870-1900

Gift within family	5.9%
Purchase within family	13.4%
Purchase from outside family	77.9%
Other (patent, etc.)	2.6%
N	(611)
Price per Acre	
Purchased from	
Father	$42.8 (100)
Other relative	$73.3 (26)
Nonrelative	$116.0 (502)

Source: Sample data.

hardware, feed, and lumber stores showed that the average amount outstanding on the books was $6,388. The median number of accounts carried in each store was 174, and account balances averaged $36.70 each. Implement dealers usually charged between 6 and 8 percent on notes, and 85 percent of all customers bought on time. Implement dealers were the only merchants who gave discounts for cash transactions. Grocery stores neither gave discounts nor charged interest on small sums because of the trouble involved. Most store credit ran for not more than six months, with interest charged on 4 percent of transactions after thirty days, 40 percent after sixty days, 25 percent after ninety days, and the remainder after longer periods. At the time of a transaction the merchant usually asked farmers when and how they planned to pay. If the bill was paid promptly, a somewhat larger amount of credit would be extended if it was wanted. If payment was slow, further transactions were made more cautiously. Provided it was agreed on beforehand, merchants would take any commodity in payment—butter, eggs, potatoes, fence posts, firewood, or railroad ties.[15]

Perhaps the most striking aspect of local credit was the businesslike way in which familial transactions were carried out. While a certain amount of this kind of borrowing never reached the courthouse, it is remarkable how many families employed the official channels for transactions within the family. There is enough evidence to suggest that intrafamilial real estate transfers were rarely completed without strings attached and, further, that most borrowing between relatives required interest payments, long before government regulations mandated the charging of interest on loans between relatives.

As table 7 shows, gifts of land between relatives were rare among those who began farming between 1870 and 1900. Perhaps this was to be expected in view of the pervasive ethic of individualism, which required sons to make careers for themselves. Yet, while children were hardly ever given land gratis, they did receive a discount in price compared to that paid by nonrelatives. Similarly, mortgage loans were also discounted between relatives when the price per acre was calculated.

There is more evidence of familial discounting when farmers sold their land at the end of their careers. Unfortunately, the numbers are small, but there are enough to confirm that relatives had an advantage when they needed to borrow money in a seller-financed transaction. In this instance the senior generation would orchestrate an intervivos transfer in which the son or daughter made a small down payment and paid the bulk of the purchase price in installments at an interest rate somewhat lower than that available from institutions or individual lenders. The same sort of mechanics were involved with seller mortgages for nonrelated buyers, but interest rates were higher. Sellers usually charged their relatives 5 percent and nonrelatives 5.2 perent, while institutions charged 5.6 percent. These seller mortgages were for the most part intervivos transfers. A good indicator of the business emphasis on familial transactions was the fact that only about 5 percent of all deeds specified support for family members as a condition of sale.

It seems that local and familial borrowing remained important sources of credit when relatively small sums were needed. With the rise in land prices after the turn of the century, the neighborhood market lost some of its importance. However, at the time of the steepest plunge, in 1920, seller mortgages were extensively employed again only to decrease in importance as the economy suffered a deflationary spiral. Few Iowa communities escaped the plunging psychology of the years immediately after World War I. In Smithfield, Fayette county, a number of the remaining pioneer Yankee families made shrewd land deals with newcomers and retired on the proceeds to the suburbs of Los Angeles, but others were not so fortunate. One seller financed the sale of three hundred acres for over fifty thousand dollars. A few years later the original owner found himself in possession of the land once more when the new owners could not keep up their payments. It was not until after World War II and the appearance of contract selling that the local seller mortgage came back into fashion.[16]

The Advent of Commercial Lending

For a number of reasons, commercial sources of funds were utilized in some communities relatively early. Such was the case in the most westerly township sampled in Plymouth County, which was settled on land owned by the Illinois Central in the 1870s. After the expiration of school fund sources, the farmers relied solely on commercial loans from institutions in Le Mars and Sioux City, from banks in eastern Iowa, or from insurance companies on the East Coast. This situation materialized because of relatively late settlement and the fact that most farmers were cattle feeders with a need for credit at regular intervals. Since community sources could not satisfy this constant demand for cash, commercial institutions filled the breach.[17]

Commercial mortgages also made inroads before the turn of the century in communities where old stock farmers sold out to immigrants. In such sales, sellers had little in common with buyers; often they did not even speak the same language. Brokers were used to attract buyers, and sellers wanted to receive the total price for the land and not be bothered with a seller mortgage. Nevertheless, it would be wrong to suggest that one method of financing dominated transactions at any particular time.

In the twentieth century the commercial credit system grew more sophisticated. A whole network of agents and middle-men working for loan companies and larger commercial banks began to serve the countryside. The business of Hugh H. Shepard of Mason City from the turn of the century to the Depression illustrates how the connections between farmers and major commercial sources were maintained. As a broker Shepard worked on a commission basis and often employed a "forwarder"—usually a bank officer in a country bank—to bring him farm business. It was the broker's task to put the farm borrower in touch with a "correspondent" in a city bank, insurance company, or other commercial lending agency. The broker had the borrower fill out the necessary forms for the loan, appraised the land, checked the title, performed a character and credit check on the farmer, and once the mortgage was approved, checked tax payments periodically and collected delinquent interest. Fees for this service were calculated as a percentage of the loan in question and reached a peak of 2.6 percent in 1919, after which they declined.[18]

Obviously such a formal, long-distance transaction was a far cry from the close confines of the neighborhood where a farmer often borrowed from a relative or friend. Because of ignorance about prevailing interest rates, borrowers were at a disadvantage and had to rely on the broker to find a equitable rate. In practice, however, although brokers did seek favorable interest rates, more important was access to a loan outlet, especially in boom times such as 1919-1920, when lenders were besieged with more applications than they could handle. Farmers affected by the boompsychology were not particularly concerned about the amount of interest they would have to pay; rather their principal concern was to get onto the bandwagon in order to make a profit from their land purchase. Loan costs were often increased because farmers did little about obtaining a needed loan until the last possible moment. Brokers, therefore, had no time to shop around for the cheapest possible outlet.

The voluminous Shepard papers give a glimpse of how the broker and his client worked over a long interval. The vast majority of cases were routine: loans were financed and paid off or extended with few ramifications. However, in the twenties and thirties many of his clients found themselves in difficulties, and Shepard fell into the role of go-between. The correspondence leaves the impression that the institutions remained aloof and businesslike in dealings with their clients but that Shepard made every effort to remain on good terms with both borrower and lender. For example, in May 1925 he wrote to the agent of the Northwest Mutual Life Insurance Company in Des Moines on behalf of a farmer named Nelson who was in some difficulty over payments on a fifteen-thousand-dollar mortgage. Shepard reminded the company of Nelson's good record in the past and hoped that they would be flexible. "The last few years," he wrote, "have been difficult ones. Mr. Nelson does not wish to appear too arbitrary in this matter, but he knows his own needs and limitations, and is too proud to ask for an extension himself."[19]

If brokers had to intervene on behalf of farmers in difficulty, they also had to be firm with those who refused to make an effort to repay debts. In the early months of 1930, Shepard had a lengthy correspondence with a widow in Riceville, Iowa, whose sons apparently had failed to make payments on a mortgage from Bankers Life. The company had begun foreclosure proceedings and Shepard urged her to protect her investment: "You can readily understand that if the Dunlay heirs do not care enough about the land to pay the delinquent interest and get the present first mortgage extended, that the Bankers Life Company could hardly be asked to extend the mortgage for its full amount without asking that some payment be made on the principal in the next five years." Shepard enclosed a stamped self-addressed envelope so that he would hear from his client by return mail. Unfortunately, the farm was foreclosed soon afterwards.[20] Brokers were required to perform a good many unpleasant tasks during the twenties and thirties. On the whole though, if the Shepard business was any guide, good brokers retained a sense of proportion in their dealings. Shepard, at any rate, never failed to exude good humor and pleasantries in his routine correspondence to all his clients.

Short-Term Borrowing

Before the middle 1920s, systematic evidence on short-term credit is not available in Iowa. Interviews, however, revealed a pattern similar to that used to finance real estate. Again, local individuals formed the main source of supply and conducted the transactions in businesslike fashion, charging interest, which was higher in the short term, on loans of a month or a week's duration. Institutional borrowing began to play a major role in short-term lending, as in real estate, only in the twentieth century. Although there is no Iowa data available, consumption and operating loans are too important to ignore. Luckily, there is a unique body of data from Michigan, 1928-1937, which to some extent fills this gap. The material comes from nine Michigan counties, seven of which had corn-belt type agriculture. In addition, since over half of almost nine hundred respondents made their living in general farming on 80-to-120 acre arms, it would seem that the material has some relevance to the corn-belt as a whole.[21]

The study charted production and consumption loans from small country banks through the worst years of the Depression. Fully 88 percent were production loans for the purchase of livestock and machinery, for labor and machine hire, or for feed and fertilizer; the remainder were for consumption—living expenses and automobile purchases, for example. A little over 50 percent of all loans were under $100, with the average amount being $400 for truck or tractor purchase, $325 for livestock, $175 for taxes, and $125 for medical expenses. Most of these loans were not secured, but if they were, the security was usually on the endorsement of a member of the family. The loans were usually for periods of between 60 and 120 days, with three-quarters of them for less than six months. Automobile and livestock loans required the longest terms, of twenty-four and fifteen months respectively.[22]

In view of the poor economy, there was a surprising liquidity to these loans. Out of 2,283 cases, just over 60 percent were liquid, in the sense of being paid in full at the maturity date of the original note. Because clients could renew or consolidate their notes, some farmers remained continuously in debt to their bankers for long periods. For instance 247 balances were outstanding at the begining of 1928; 33 percent were repaid at some time during the year, but another third remained outstanding until 1936. These were probably representative of "tide-over" loans. Because they were chronically short of cash, most farmers sought to borrow as much as they could whenever they went to the bank in order to temporarily alleviate their cash-flow problems.[23]

The Michigan consumption and production loan data reveals the same cyclical variations found in the Story County real estate mortgages. Farmers would seek a large volume of credit in relatively prosperous years, only to find that they had difficulty with repayment when the economy went sour, as it did in 1928. By the final quarter of 1929, farmers began avoiding their obligations. By 1930 and 1931, as the Depression deepened, although lenders and borrowers were more cautious and selective, loans proved increasingly difficult to collect. Only in 1932, when prices fell to all-time lows did liquidity show an upturn—mainly through "forced liquidation," that is, from the liquidation of a debtor's assets. In 1933, 95 percent of all loans were paid in twelve months, indicating both that the Depression had bottomed out and that bankers were being very cautious about lending.[24]

Among other things, this data points to the importance of credit in almost every phase of farming. Borrowing was seasonal—in the spring for planting and in the winter for livestock purchases and upkeep. Above all, amounts were small, while the lack of security for loans showed that, despite the economic situation, local institutions relied on their customers' good faith to repay loans eventually.

Lending in the Central Valley

Farm credit mechanisms in California had many of the same elements as in Iowa. Where small farms predominated, farmers behaved like their peers in the Midwest, lending money to children and neighbors. Small local institutions did most of the lending in the early years of the century, but most did not survive the Depression. California, however, initiated branch banking in rural areas, and lenders like the Bank of Italy began to buy local institutions in the twenties. The Bank of Italy, or the Bank of America, as it came to be known, remembered its origins and took particular care to integrate itself with its local farm clientele in any community it entered.

By way of introduction to the credit situation in the Central Valley, it is convenient to survey transactions in a random collection of tracts from the turn of the century until the early sixties. In the Central Valley, the type of farming activity often determined the amount of borrowing and the kind of credit. The fruit rancher's needs were unlike the grape grower's, and dry wheat farming had less need for financing than did dairy.

A trace through the credit transactions in the southeastern quarter (160 acres) of one township in Kings County, from the time Southern Pacific sold the first land to the 1960s, begins to reveal the pattern. From 1879 to 1914, seven mortgages were issued; two were from the local Hanford bank, and the remainder were either individual or seller mortgages. Until 1890 the land was dry farmed for wheat, but during the last decade of the nineteenth century, part of it was split up into twenty-acre parcels for grapes. By the turn of the century, all the land was in the hands of two families that would own it until the sixties. Both families bought their land by means of seller mortgages from a Monterey widow. In one of these transactions, the Vaughans borrowed $2,275 at 8 percent in 1910 for the east half of the quarter section. Four years later, in an intrafamily transfer, a son borrowed $1,000 from his father at no interest, agreeing to pay back the loan within five years. Apart from a Sun Maid raisin contract on one of the parcels, there was an unusually quiet period in the twenties and thirties, until 1939 when a renter obtained a crop mortgage for four hundred dollars at 4.5 percent from the forerunner of the Farmer's Home Administration. If mortgage data is any guide, it would seem that this tract was put into cotton for the first time after the war. In 1949 season a renter obtained a crop mortgages from Western Cotton Oil. From then on, throughout the fifties and sixties, the Anderson Clayton Company served the needs of various tenants with cotton production loans.[25]

A different pattern developed on the northeastern quarter of another township in Kings County, south of Hanford. This tract, though sold by the railroad in 1881, did not become productive until cotton was introduced in the early thirties. For instance, in 1904 the Mercantile Trust Company lent William Pearce $2,200 to raise grain on 320 acres, which included this quarter. But until 1930, when a new owner obtained a crop mortgage from California Cotton Credit Corporation, there appears to have been only sporadic farming activity on this ground. In the middle of the thirties the land was bought by an insurance company, which sold it to the present owners, cotton growers, who turned to the Visalia Production Credit Association for loans in the late thirties and early forties. One renter received three different loans from the Bank of America in the fifties. Until the 1960s, however, most credit was supplied by large ginning companies such as Anderson Clayton.[26]

In Tulare County, south of Dinuba, the intense cultivation of raisin grapes got under way must after World War I. Sun Maid had contracts with six growers in the northwestern quarter of one township in July 1923. The majority of the owners were Mennonites, who expanded the production of raisins from 1923 to 1928 with production loans from the Bank of Italy, First Los Angeles, and the Bank of San Jose. However, a downturn in the spring of 1929 caused some of the borrowers to default, and until the Land Bank began to lend money in 1934 for the consolidation of ownership, there was little further borrowing activity. Indeed, the shock of the Depression was such that until the sixties, borrowing was done with extreme caution. Mostly this was in the form of modest loans from the Federal Land Bank or crop mortgages from Anderson Clayton or Western Cotton Oil. In the 1960s, when the Mennonites were preparing intergenerational transfers, two families employed classic midwestern techniques. Both sold their sons tracts of land through seller mortgages. Each son made a small down payment, and the parents used the mortgage payments as an annuity.[27]

One final Tulare example comes from the citrus belt on the edge of the Sierra foothills. The southwest quarter of one section was bought in the late 1890s by A.C. Merryman, who incorporated the business three years later. After Merryman sold out in the late twenties, the orchards changed hands several times. In 1930, for instance, there were no less than three crop mortgages on the property, two of them over thirty thousand dollars each. In the following year the operator went bankrupt. For the remainder of the thirties and forties the orange groves were owned and operated by the Western Fruit corporation, which in 1939 borrowed seventy thousand dollars from City Trust and Savings of Los Angeles for operating expenses. In 1957 a private individual again purchased the property. Oranges had become, for a time at least, a useful tax write-off. The new owner was a wealthy retiree from Los Angeles, who in 1965 borrowed half a million dollars from the Prudential Insurance Company.[28]

Probably the most striking aspect of Central Valley ranching was the early need for operating loans, which did not occur in the Midwest to the same extent until later. Farmers in the Central Valley looked for credit where they could find it, whether from local banks, branch banks, or large commercial institutions on the coast. Insurance companies were not as active among California farmers as they were in Iowa. However, federal sources of funding made an earlier impact in the Central Valley. Land Bank loans and credit from Production Credit Associations were increasingly used after the middle thirties.

Small-Farm Borrowing

To analyze credit mechanisms among small farmers in the Kings delta I randomly selected mortgages from the years 1930 and 1946. As in the Midwest, small farmers used short-term loans quite regularly. Local banks at Laton, Lemoore, and Hardwick—all tiny trading centers—made many such loans. Most of the transactions were for less than two thousand dollars, interest rates of 7 to 8 percent, and terms were usually for quarterly payments of the principal over one year. The smaller the loan, the higher the interest rate and the less time there was to pay. For instance, M.P. Dutra borrowed four hundred dollars from the Hardwick Bank at 8 percent in 1930, using his ten cows and a bull as security. His payments were twenty-five dollars a month.[29] In 1930 some dairy farmers borrowed operating money on a regular basis. Most of this activity was for quite small sums of money, and local banks and individuals in equal numbers supplied the necessary funds. Metropolitan bank involvement in these kinds of transactions was rare. Indeed, the only such loan was to an Anglo fruit farmer for five hundred dollars at 8 percent with eleven months to pay. His farm was one of the few that had a tractor at a time when mechanization was minimal on small California farms.[30]

The most noticable change seventeen years later, in 1946, was the lower interest rate of around 5 percent on government loans, with commercial credit somewhat higher. Not only had federal programs such as the Farmer's Home Administration been initiated but both the Land Bank and Production Credit were competing with commercial lenders for business. Because of inflation and the greater scale of operating, loans were larger. When Jess Williamson borrowed $6,350 from the Bank of America in 1946 he had two years in which to repay the loan, but his interest was high at 7 percent. Security in this case was a herd of sixty cows, a John Deere tractor, and a pickup truck. Federal credit agencies offered lower rates and better terms than commercial lenders. For example, Clarence Brady had four years in which to pay back twenty-five hundred dollars at 5 percent from the Farmer's Home Administration. At the same time, individual loans were still being made after the war at quite liberal terms. On one occasion one farmer lent eight thousand dollars to another at 4.5 percent interest, giving him twenty-eight months to repay the loan in three-hundred-dollar installments.[31] However, as these data show, even among small farmers, agricultural credit had grown so much that the prewar system of local banks and neighborhood borrowing was almost a thing of the past.

Production Credit in Cotton

By far the most important development in the growth of credit in Tulare and Kings counties was the introduction of production credit for growing cotton from commercial processors such as Anderson Clayton and Western Cotton Oil. Commercial ginners and processors came to the Central Valley from Texas and Arizona in the early twenties. They believed that irrigated cotton had great potential in the nonhumid Central Valley and that growers, once initiated to cotton, could guarantee them a high-quality product. By 1925 there were eight gins in the valley, three of them in Kings and Tulare counties, at Corcoran, Westhaven, and Porterville. All the ginning companies had had experience with cotton production in the South, and they began to experiment with similar lending techniques in California. Their opportunity to integrate finance and production came about because local banks were unwilling to risk a new and untried crop. For the first fifteen years of cotton production in the Central Valley, growers and commercial ginners had a very close relationship. Although there were occasional grumblings about a system that had originated in the crop-lien system of the South and was notorious for its unfairness to farmers there, a feeling of mutual dependence seemed to hold both parties together until other arrangements surfaced just before World War II.[32]

Under the new system, ginners lent money to growers with the understanding that any cotton produced would be sold to the lender for ginning. As a general rule, loans did not exceed the amount a farmer could be expected to repay from one year's crop. At the same time, loans were often designed to take care of more than just growing and harvesting costs. In some instances, they would cover machine purchases, winter crops, and the rental of more land. The cotton producers were professional lenders like banks and government lending agencies, and like them, ginners sat down with farmers to work out their cash requirements for the year. Once per-acre growing costs were estimated, management would approve a sum to cover production costs. Farmers would draw this money in monthly installments at a certain rate of interest or whenever they needed it. Throughout the year, financing was required for land preparation, seed and fertilizer purchase, planting, chopping, power charges for irrigation, and a myriad of other expenses involved in growing a cotton crop. As harvest neared, producers would provide weekly advances, sometimes arranged through an additional loan, to meet the picking payroll. Once the cotton was delivered to the gin and the loan and interest were paid, the rancher could figure out his profit. To secure such loans, ranchers mortgaged not only the crop but also land and buildings. Larger loans involving thousands if not millions of dollars, might require a second mortgage as well. In each case ginners insisted that all cotton and cottonseed be delivered at harvest. It was a way to guarantee a definite volume to the gin and provide for orderly planning, while giving the grower a chance to make a living.[33]

By World War II, commercial banks had finally begun to support the cotton industry. The larger institutions, such as Bank of America, financed the cotton producers, and they in turn lent money to growers. It was an example of the interrelatedness of agribusiness, where the largest growers served as directors of cotton processors. Like other marketing and producer organizations that came into existence in the first decades of the century, the cotton ginners forced ranchers to make their operations more efficient, to pay more attention to how the operations were run, and to support the notion that farmers were engaged in small businesses. In the early years of cotton production and when machine picking was being introduced, such dependent relationships with commercial companies were probably beneficial. Except for fieldwork, growers had few concerns; even bookkeeping was taken care of by the ginner. Inevitably, however, as growers became more secure with the crop, a certain proportion preferred to look after their own interests and seek more flexible arrangements elsewhere. Thus, in the fifties cooperative gins began to open, and these institutions took much of the business away from the commercial firms.

Government Loan Programs

The giant Farm Credit System had its origins in the New Deal. Along with the rejuvenated Land Bank—originally founded in 1916—the Production Credit Association, and the Bank for Cooperatives grew out of a desire on the part of the Roosevelt administration to assist farmers with their credit problems. Much of the effort in the early years was given to refinancing farmers who had lost or were losing their farms as a result of the downturn. Another important agency emerged a little later with the passage of the Bankhead-Jones Tenant Purchase Act of 1937, a response to dissatisfaction over the continued high tenancy rates in the country. The program initially was administered by the Farm Security Administration, but in 1946 the Farmer's Home Administration took over its duties.[34]

Fortunately there are a number of studies that assess the performance of these agencies on farm credit in the Midwest. One deals with Land Bank borrowing in the four north-central Iowa counties of Franklin, Hamilton, Hardin, and Wright from the founding of the Land Bank in 1916 until 1947. Throughout these counties there were 2,806 Land Bank loans closed throughout the period, of which only 697 were made before 1933, 1,326 between 1933 and 1936, and 783 from 1937 to 1947. As these counties possessed some of the best land in the state, and a Land Bank loan was still something of a rarity, it was not surprising that only 10 percent of the loans closed between 1917 and 1922 were foreclosed, 13.7 percent between 1923 and 1927, and none between 1928 and 1932. Even in the depths of the Depression, only 4 of 1,326 refinancing loans closed between 1933 and 1936 were foreclosed. A close look at the forbearance program reveals why. Not only were rates of interest reduced to 3.5 percent, but payment of principal was waived for a period of five years between 1933 and 1938, provided interest, insurance, and taxes were paid. In 1940 the Federal Land Bank of Omaha instituted a variable and suspended plan of its own. As long as a borrower complied with the terms of the mortgage—that is, to pay taxes, insurance, etc.—payments on that loan could be made by (1) "delivery of a fixed share of all crops when harvested, to be sold at the option of the mortgagee and applied upon the mortgage or sales contract debt, or payment of a fixed percentage of annual gross income; and (2) periodic fixed cash payments."[35]

Another way to look at this data is to inquire when such foreclosures as there were actually took place. Indebtedness was not usually as heavy on farms with Land Bank loans as on other farms. Apparently in the four-county area, Land Bank loans survived virtually unscathed during the twenties, while loans from other lenders were breaking down. It was not until the thirties, when prices dropped very low, that Land Bank loans closed in 1917-1932 began to fail. A heavy debt load was the most common reason for foreclosure. Often these debts were junior real estate or chattel mortgages. In a number of instances a second mortgage was held by a bank that closed, forcing the Land Bank to foreclose to protect its investment.[36]

On farms with better land, it was significant that foreclosures resulted in slight gains for the agency. Even on the poorer farms, where 22 percent of all loans were foreclosed, there was only a $3.50 loss per $100 loaned. Finally, given the multi-billion losses suffered by the Land Bank during the eighties, it is interesting to note that in a previous period of extreme economic stress, a policy of forbearance obviously had an impact on the incidence of repayment and contributed to its success. When the thirty-five-year loans were classified by farm type, most were ahead of schedule. For example, as many as 64 percent of all loans issued between 1917 and 1932 to the lowest-rated farms were repaid by January 1, 1948, several years ahead of schedule. And the rate for the best farms was 83 percent.[37]

The performance of the Farmer's Home Administration and its predecessor, the Farm Security Administration, was also studied in Iowa from 1938 to 1946. In the early years, the agency had an extremely cautious lending policy and vetted borrowers stringently. During those years, a total of 921 loans were made in the whole state. Loan limits were set at twelve thousand dollars in 1946 and interest rates at 3.5 percent. No down payment was required, and veterans received a preference inclosings. In 1946, of a total of 5,626 applications received in Des Moines, only 194 were approved.[38]

In a sample studied in 1946, virtually all borrowers were ahead of schedule. Some borrowers used their opportunity to maximum advantage to get ahead in the high-income war years; others were less ambitious. For instance, one farmer with a sixty-five-hundred-dollar loan was so intent on paying off his obligation that he held expenses to a minimum in order to do it. Rather than expand production, he put everything he earned into accelerated payment. In all, 28 percent of the sample had paid off their thirty-five-year loans by 1946, and most of them were totally dependent on farm income to do it. The evidence suggests that the restrictiveness of the program made it only marginally successful in the early years. At the same time, those fortunate enough to take part appreciated their opportunity. Numerous letters in the agency files testify to the hope and confidence that the program gave its small number of participants.[39]

Evidence from an early sample of Minnesota Production Credit clients, reveals much the same pattern. Production Credit Associations were concerned with short-term credit for operating loans. In 1948 these were usually modest, with $2,450 an average-sized loan. Money was provided for operating as well as machinery purchases, with smaller amounts for renewals. The record of repayment seems to have been reasonable. Poultry and grain farmers had the best performance, and dairy farmers the worst, apparently, because milk prices fell unexpectedly in the summer of 1948, creating repayment difficulties. Unlike commercial institutions, the PCA required budgeting procedures. At the time of application, farmers had to spell out in detail how the advance would be used and in what time period.[40]

Farm lending remained conservatively oriented for three decades after World War II. Inflation was relatively low; interest rates on long-term Land Bank loans were pegged at around 5 percent; there was considerable movement out of agriculture; and the memory of the Depression tended to disuade farm families from taking on large amounts of debt.[41] National data show that the ratio of debts to assets declined from 18.9 percent in 1940 to 8.7 percent in 1951, although there was a modest increase to 11.9 percent in 1960. In comparison to the middle eighties, when only one-third of all farmers had debt-to-asset ratios of around 10 percent, debt assumption was of very modest proportions in those years.

A decade later national figures from the Federal Land Bank documented "the excellent repayment record" of farmers. In 1967, 97.3 percent of all borrowers had met all their scheduled installments, and this record had improved by two percentage points from that of 1957. Only twelve farms in the whole country had reverted to the Land Bank by default in 1966, and all were disposed of in the same year. In Iowa only 1.1 percent or 225, of all outstanding loans were delinquent as of June 1967; in California 310, or 2.6 percent, were delinquent.[42] One feature of these years, however, was the increased role of the federal credit system in both short- and long-term lending. An overview of the California farm finance system showed that in farm real estate loan holdings from 1960 to 1967, the share of the Federal Land Bank had grown steadily, increasing from 13 percent of the market in 1960 to 23.7

percent in 1975. Life insurance companies also increased their share from 13.3 percent to 22.7 percent Commercial banks fluctuated between 6 percent and 8 percent, and individuals, though still the largest category, declined from 66.7 to 46.6 percent in 1975. A similar trend occurred with operating loans. Production Credit Associations in California doubled their share of business, from 14 percent in 1967 to 29.1 percent in 1976, while commercial banks—still by far the largest suppliers of farm operating credit—dropped form 83.4 percent to 69.2 percent in 1976.[43]

Farming the Government, 1945-1986

Farm programs are as controversial as they are complex. With the passage of time they have become phenomenally expensive for taxpayers as well. They are of interest here because of their income-producing capacity for corn, cotton, and dairy farmers. Furthermore, a certain percentage of support programs come in the form of loans, which farmers can either pay back to the government or retain, depending on the price of a given commodity. Agricultural policy was traditionally controlled in Congress by a coalition of conservative southern Democrats and midwestern Republicans. With protracted struggles and compromises the contending groups attempted to forge agricultural policy with the passage of a farm bill every five years. Two criteria, which were at cross-purposes with each other, governed farm programs after 1945: the desire to reduce excess production and the need to support the industry in a period of instability.

The major achievement of the farm program in the years 1945-1981 was the security it gave family farmers. The fear that agriculture would go through a slump after World War II, as it had done after World War I, was averted. Likewise, the period of excessive capacity during the fifties and sixties did not result in the decline of agricultural prices. This stability in farm prices and incomes was also translated into low consumer food prices and an abundance that was used to feed hungry people round the world.

If they were judged solely for their efficiency, commodity programs had shortcomings, for an emphasis on social rather than economic criteria resulted in a certain amount of waste. An unwillingness to impose strict production controls and a tendency when they were imposed to permit cross-compliance, which allowed a farmer to shift production from one crop to another, caused further surpluses. In addition, acreage controls were a "weak and slippery" form of control. Farmers tended to place their poorest ground in the program and, with new and improved technologies, increased output on the remainder.

As a result, total output did not decrease. Moreover, controls through land retirement encouraged the substitution of machinery, chemical fertilizers, and pesticides for labor. The programs encouraged an increase in the flow of off-farm inputs to agriculture and the loss of employment for those dependent on farm jobs. The programs tended to benefit the larger farmer rather than the medium-sized or small operator.[44]

By the middle sixties, the crop programs had three components: production controls; so-called deficiency payments, which were direct cash payments to farmers; and Commodity Credit Corporation loans. Deficiency payments, pegged to preset "target prices," assured farmers of the income they would have received if market prices were at target levels. Farmers received the difference between the target price and the loan rate. If the loan rate for corn was $3 a bushel and the target price was $3.50, but the market price fell below $3, the farmer would sell his crop to the government for $3 a bushel, and keep the 50 cents as a cash payment. In order to qualify for deficiency payments and CCC loans, farmers had to agree to set aside some of their land. The object was to pay for some of the program's cost by shifting it from the taxpayer to the consumer and, by cutting back on production, to raise prices. The Commodity Credit Corporation loans worked differently. The farmer borrowed the equivalent of the value of the crop and used the crop as collateral. If the market price turned out to be higher than the loan rate, the farmer sold the crop and satisfied the loan. However, if the market price was below the loan rate, the farmer was able to turn over the crop to the CCC and retain the loan.[45]

The objective of the grain storage programs in 1933 was to prevent year-to-year fluctuations in price. Over time, however, prices were stabilized upward and loan rates were set above "average-weather-crop" levels. This policy raised the level of prices, stimulated production, and reduced consumption, leading to wholesale surpluses. Both World War II and the Korean War bailed out the programs at critical stages, but after 1952 good weather and technological advances caused CCC stocks to increase. In an attempt to solve the problem of overcapacity, the loan rate for corn was dropped from 90 percent of parity to 65 percent in 1960. Two years earlier corn producers were offered the choice of higher price supports or greater acreage restrictions. They voted for the latter, and surpluses continued to accumulate.

Surpluses were expensive to store. Between 1932 and 1960 costs amounted to almost $20 billion, of which only part went to farmers. In 1958 almost a third of the cost of the corn program was for storage and handling. Because of cross-compliance, most programs were not effective production controls. In Iowa in the fifties only 42 percent of farmers complied with corn allotments because many fed corn to their livestock. Those that did comply also grew soybeans and oats. By and large, reductions in corn acreage by those in the program were offset by increases made by farmers who did not comply.[46]

If farm programs were expensive in the fifties and sixties, the average farmer hardly grew rich on them. For example, in 1948 while $257 million was paid out in programs, the average individual farm payment amounted to only $44. Ten years later just over a billion dollars were spent on all programs except dairy, and individual payments averaged $257. By 1968 $3.4 billion was spent, and a single operation averaged $1,127. In 1968, 65 percent of those in the program received payments of less than a thousand dollars; at the same time 67 percent of the funds went to 33 percent of the producers in payments ranging from a thousand to fifteen thousand dollars.[47] In 1969 approximately one-half of all net income from farming came through farm programs, whether on large farms with over forty thousand dollars in sales or small farms with less than twenty-five thousand dollars, reflecting the support objectives of the programs. On dairy farms a study of support programs done in 1963 indicated that gross income would have been 16 percent less and net income 43 percent less if milk price supports were abandoned.[48]

In contrast to the modest expenditures in the thirty years after World War II, the eighties saw a complete reversal. Aggregate figures from 1980 to 1984 reveal that Iowa alone received almost $2 billion in this period, with almost $1 billion coming in the payment-in-kind year of 1983. California farmers, who consistently espoused the rhetoric of the free market, received almost $900 million in these years. In 1985 the average crop payment nationally was $24,000, and 60 percent of the money went to medium-sized operations between $100,000 and $500,000 in earnings.[49] Crop payments in 1986 were as high as $25.8 billion at a time when $7 billion went to the space program, $9 billion to Aid to Families with Dependent Children, and $229 billion to defense. In California in 1980 only $300,000 was paid out to the cotton program when "free-market" conditions prevailed. By 1986 virtually all cotton farmers in the southern San Joaquin depended on the program to survive the downturn. With cotton payments set at eighty cents a pound and the market price half that, farmers could not afford to be free-marketeers. Many cotton farmers were resorting to a practice called sidestepping, whereby a family would set up partnerships with family members and business associates to sidestep the fifty-thousand-dollar limitation in the cotton program in order to garner the maximum number of fifty-thousand-dollar payments. In Iowa such practices were less blatant, but the Des Moines *Register* published the names of 990 farmers who received over thirty thousand dollars in the

1986 farm program. The vast majority of these operations were paid between seventy and ninety thousand dollars.[50]

As was common throughout the history of support programs, larger operators tended to benefit more than small ones. Nowhere was this better illustrated than in the 1986 dairy herd buy-out program, which was designed to cut milk production by eliminating dairy herds. In Kings County sixteen dairies received the munificent total of $17,682,479 in buy-out payments. A single two-thousand-cow herd was bought for $8 million. No wonder some categorized the farm programs of the eighties as the "new farm aid."

In short, there was a paradox about the government's strategy for agriculture in the eighties that often placed programs in conflict with each other. For example, on the San Joaquin's westside a cotton farmer received $134,000 in 1986 through the Federal subsidy program designed to make him plant less cotton. At the same time, another branch of government gave him an estimated $130,000 discount in water deliveries. "We have," said a critic of the California agricultural interests, "taxpayer-subsidized water projects delivering taxpayer-subsidized water to farmers, who use it to produce subsidized and surplus crops."[51]

The Long, Slow Boom

The long inflationary spiral after 1966 dramatically changed attitudes towards credit both on the farm and off. It is important to make the point that farm behavior in a period of inflation was no different from that of other sectors. Agricultural lenders treated the inflationary boom in much the same way as did their colleagues in two other areas that were expanding at this time: the oil industry and overseas lending. Their expansion was economically rational given the circumstances in which they were operating. Agriculture, after all, was big business with historical record of ability to repay loans from crop and livestock income. While individual farms had their ups and downs, farm income was relatively stable. Farmers, after all, owned land, an extremely valuable nonrenewable resource that enabled them to refinance short-term operating losses.[52]

Farmers who had made money in the seventies needed to invest their returns somewhere. And given the signals for expansion that were coming from all sides, it was hardly surprising that investments in land, buildings, and machinery received greater attention from farm families than more liquid investments. Lenders, not unnaturally, were pleased that their customers were doing well and were more than

Table 8. Distribution of Real Estate Mortgages, 1974-1984 (millions of dollars)

	1974	1975	1976	1977	1978	1979	1980	1981	1982	1983	1984
					Iowa						
Land Bank	644	793	1,020	1,254	1,522	1,845	2,378	2,919	3,228	3,296	3,261
FmHA	146	152	156	166	175	304	345	397	413	440	483
Life Insurance	489	479	492	628	747	846	948	988	1,023	986	965
All banks	203	234	284	330	356	358	339	321	343	420	512
Individuals	1,689	1,889	2,108	2,432	2,748	3,305	3,577	3,766	3,793	3,831	3,544
					California						
Land Bank	929	1,039	1,172	1,336	1,557	1,967	2,422	3,095	3,559	3,852	4,033
FmHA	28	25	31	36	39	114	140	211	195	211	229
Life Insurance	885	996	1,108	1,296	1,549	1,765	1,810	1,871	1,872	1,991	1,948
All banks	248	204	203	246	322	398	483	538	532	623	687
Individuals	1,463	1,555	1,647	1,804	1,933	2,203	2,385	2,511	2,529	2,559	2,363

Source: USDA, *Economic Indicators of the Farm Sector,* 220-40.

willing to lend them money for further expansion. Tables 8 and 9 graphically show these increases in both real estate and operating loans in Iowa and California from 1974 to 1984.

It has been suggested that competition between lenders was a factor in creating the booms. It was easy for a customer to go down the road to a rival bank or to the Farm Credit System if a loan request was refused. In order to keep business, lenders waived some of the rules, paid less attention to the repayment capacity of the operation, and placed too much emphasis on the equity position of the farmer. Leveraging (that is, using cheap money to purchase land to take advantage of rising land values) became a respectable strategy for both lenders and farmers to follow, and too often little effort was made to "pencil out" the cash flow of a purchase. The tax system, which encouraged end-of-year purchases to reduce tax liabilities, also was abused. Finally, bank deregulation in the early seventies shook up the industry and had a profound effect on the way business was conducted.

Perhaps if lenders had wanted to stifle the boom, they could have worked together to do so, but a number of factors were against a dampening of liberal credit. First, there was the competition for business; second, by the seventies most of the personnel who had experienced the Depression had retired, and the careful lending practices of that era were forgotten; and third, lenders, unlike other professionals, such as engineers, lawyers, or doctors, did not have to go through professional training. Standards, therefore, were not as high as they might have been.[53]

As the inflationary cycle gathered momentum, the value system of farming changed. Neighbors became very competitive as they bid against one another for land, and the boom killed old neighborhood ethic of sharing labor and machinery. In the drive for expansion, families bought all the equipment they needed to be self-sufficient at crucial times in the year.

The extremely high inflation of 1979, 1980, and 1981 brought about speculation in the Midwest often by syndicates of nonfarm investors. Unfortunately for them, and especially for those from whom they bought contracts, the economy went sour before any of the transactions were completed, leaving a trail of bad debts and broken elderly farmers. According to the Minnesota agriculture commissioner who investigated land speculation in the southern part of his state and adjoining counties in Iowa, these groups seemed to have a foolproof plan. "They never negotiated on price," he said.

> They negotiated on terms—the amount down, the interest rate, the annual payments. My own feeling is, these guys pushed up land prices deliberately. They knew what they were doing.... They'd take a farmer's first asking price, but on their terms of low down payment, low interest rate, low annual pay-

Table 9. Distribution of Non-Real-Estate Farm Loans, 1974-1984 (millions of dollars)

	1974	1975	1976	1977	1978	1979	1980	1981	1982	1983	1984
					Iowa						
All banks	1,705	2,032	2,423	2,784	3,030	3,529	3,361	3,323	3,766	3,842	3,874
PCA	410	499	607	686	779	1,068	1,149	1,127	1,114	1,071	886
FmHA	47	101	94	119	247	314	405	444	442	452	593
Individual	581	701	825	996	1,061	1,191	1,265	1,263	1,350	1,192	1,118
CCC	38	47	99	334	612	611	742	1,107	2,036	1,247	1,063
					California						
All banks	1,584	1,614	1,860	2,020	2,292	2,655	2,848	3,345	3,763	4,136	3,935
PCA	566	670	827	1,012	1,204	1,350	1,600	1,829	1,939	2,071	2,206
FmHA	8	11	10	28	125	472	555	594	588	592	694
Individual	667	759	864	1,040	1,198	1,361	1,455	1,579	1,672	1,624	1,585
CCC	6	10	14	27	87	97	22	25	161	73	57

Source: USDA, *Economic Indicators of the Farm Sector,* 220-40.

ments and a balloon five or ten years later. We feel they never intended to pay the balloons. The idea was to concentrate on purchasing, and start prices up, and then you get out, you can make a lot of money. . . . whether it's intentional or not, it's a strategy based on inflationary land values.[54]

Just over the Iowa line in Fillmore County, Minnesota, roughly ten thousand acres were bought up by speculators at this time. Rather than work the land themselves, the owners hired custom operators, who tore up the land, destroying waterways and terraces in order to plow hillsides to plant corn. In 1958 in this area there were 196 family farms, but with their purchases the outsiders had eliminated this very stable community. While the behavior of the absentee owners in Fillmore was especially notorious, several counties in Iowa—often those with mediocre soil and a history of erosion—saw similar practices.[55]

Another measure of the changed climate was the common practice by the seventies of using short-term operating funds to make long-term mortgage payments. One side of the Farm Credit System, PCA, lent borrowers funds so they could make their mortgage payments to the Land Bank. Similarly, the Farmer's Home Administration also joined the rush to attract new customers. The agency underwent a huge expansion in the late seventies, especially with disaster loans. The "lender of last resort" took on accounts that in earlier times it would have not entertained, and its lending policies therefore tended to encourage risk-taking. Thus by the middle eighties in both the federal credit system and the FmHA, bad debt had a tendency to follow bad debt leading to disastrous consequences when the bubble finally burst in both institutions in 1984-1985.[56]

Conclusion

Farm families need financial intermediaries—local savings institutions, elevators, cotton gins, the Farm Credit System, FmHA, local and metropolitan banks, relatives, and neighbors—to provide funds over long and short periods in order to operate their businesses. No family has the same borrowing and saving requirements as any other. Operations and family composition are so different. Thus, a dairy farmer would not have the borrowing needs of a grain farmer, whose cash flow would be limited throughout the year; a farmer with a son who wanted to farm would require long-term financing to purchase land, whereas another without heirs might want to sell land through a seller mortgage. The modernization of credit mechanisms—the upgrading of the Farm Credit System, the advent of production credit in cotton, and the increased importance of chain banking—failed to dislodge the important role in the individual played in long-term lending for real estate. Throughout the twentieth century, except in times of high inflation, seller mortgages and contracts from relatives and neighbors continued to play a useful part in borrowing and saving in the farm community.

Since the thirties the impact of the Land Bank of the Farm Credit System for long-term real estate mortgages was considerable. Not only were Depression-era interest rates far lower than those in the nineteenth and early twentieth centuries, but terms were longer—often thirty years, compared to five or ten, with balloon payments at termination, from institutions before World War I. In addition, the Depression conditions called for forbearance on loans if the mortgagee had difficulty paying principal and interest. Forbearance allowed the Depression generation to ride out hard times and survive until prices for commodities grew better. This generation of farmers was generally exceedingly cautious in its credit behavior because of its experience in the twenties and thirties.

But the next generation, the one that started farming in the fifties was not so fortunate. For them, the changes in the structure of agriculture—the increased size of farms and number of inputs—required more credit than ever before. In California, medium-sized operators routinely borrowed hundreds of thousands of dollars each year in the seventies. Even in the Midwest, large sums were needed to put in crops and to purchase and feed livestock. The change in the attitude towards credit among both farmers and lenders had a number of implications for the middle eighties. Credit had become a ticking time bomb.

6 Family

In Iowa and the Central Valley, some families and communities have survived and prospered despite the disruptions that have beset agriculture in this century. So far, the many challenges farm families have faced have been discussed in general terms. Now it is time to examine the record in greater detail to try and understand why, in an occupation characterized by massive out-migration, contraction, and concentration, certain families managed to remain on the land.

Traditionally, corn-belt farming has favored the family-type or tenant-type farm. Here, owners or tenants have done the managing, and they or their families have supplied the labor. The seasonal nature of agriculture in the corn-belt worked against the permanent employment of a large labor force, for grain farming is characterized in the Midwest by short bursts of intensive labor interspersed with slack periods. Livestock raising, and especially dairying, on the other hand, required year-round commitment to an exacting daily round of chores, which were, and still are, best supplied by a captive family labor force. The family, with the ability to utilize quite young children, spouses, and other relatives, allowed an operation to overcome the built-in inefficiencies of slack and active work time. The farm family, therefore, on the family or tenant farm, was unique in that it allowed economic and social roles to coalesce. The father and husband was owner, manager, and laborer; the wife also played an important economic role. In addition to managing the home, certain outside tasks were traditionally hers as well. Today, when a farm wife's off-farm income can help a business to stay afloat, her contribution is even more important. Most farm operators were dependent at crucial times on the labor of children, although the mechanization of farming freed them from much of the traditional drudgery.[1]

In the Central Valley, the scale and type of operation determined whether or not a family head worked as a traditional laborer and manager or hired employees for most of the work. This dichotomy existed almost from the start of settlement, so that the larger-than-family-farm, which was owner managed but used considerable outside hired labor, stood side by side with the smaller family-type farm. On a small fruit ranch a major portion of the labor was often performed by someone other than the operator. For this reason the classification of farms into one category or another is somewhat arbitrary.[2] In the contemporary Central Valley context, the family-type farm is one that uses "little hired labor," according to one criterion, "approximately" sixty days of work by those hired outside the family.[3] After World War II, the advent of the closely held family corporation further systematized the staff and line organization of larger operations. Again, it is difficult to differentiate between the larger-than-family farms, and those categorized as industrial-type organizations. On the lake bottom, most of the big ranches were family managed but also employed salaried personnel in key positions. In addition, the labor force in such an operation could be several hundred strong. Were these operations "family farms"? By the standards of the corn-belt, they obviously were not. On the other hand, they were certainly family businesses and encountered many of the vicissitudes associated with family ownership and management.[4]

At the same time, the more rigid arrangement of work on the Central Valley farm never prevented family members from doing many of the same tasks that their midwestern peers routinely performed. Even though a Central Valley ranch might employ irrigators, tractor drivers, milkers, office workers, or other specialized personnel, that is not to say that family members did not fill in at such tasks or serve apprenticeships in line jobs before assuming a supervisory role. Before World War II, if the absentee, and nonfarming landowners can be ignored, the Central Valley farm family acted much like its midwestern counterpart. Portuguese dairy farmers, for example, were exactly like German Catholic dairy farmers in the Midwest in maximizing family labor. And even lake bottom farmers prided themselves on their pioneer origins, delighting to point out that in the early days they worked in the fields with their hands, while their wives cooked for the work force.[5]

Given the difficulties of orchestrating and blending together a family's talents throughout the developmental and family cycle, it is not altogether surprising that farm family continuity on the same farm and in the same community over several generations was uncommon. The prevailing ideology on old-stock American farms emphasized individual achievement, and each generation was expected to make its own way, whether in farming or outside it. In contrast, European immigrants who settled the corn belt and Central Valley came from an entirely different tradition. Although landownership was rare, for most families came from tenant or peasant backgrounds, leases often

Table 10. Characteristics of Familism and Individualism in Farming

Familism	Individualism
Family helps members establish themselves in farming	Children fend for themselves when they begin farming
Families work together continually	Exchange work with no one or with non-relatives
All children in family become farmers	Siblings often enter nonfarm work.
Continuity of home place, which is an emotional center of family	Indifference to their association with farm
Deliberate policy of doing business with family and solving problems within family	Independent action, wary of doing business with relatives

Source: Rohwer, *Family Factors in Tenure Experience,* 829-30.

were passed down from father to son. In return for the privilege of working the land, families had to meet certain obligations that tended to guarantee intergenerational continuity. For this reason, European rural social structure had a stability that was never characteristic of American rural society.[6]

As a result, European ethnic groups who settled the corn belt and Central Valley behaved differently from their native-born neighbors, giving the family priority over the individual. Untiring industry and frugality produced a conservative ideology among Germans in Iowa and Portuguese in the Central Valley, which resulted in stable family operations. Over time intermarriage with neighborhood families gave further access to landownership and increased family holdings. Traditional institutions such as the church, parochial school, and social club further cemented family and neighborhood in both areas.[7]

In the rural Midwest it is still possible to identify communities that have been pockets of stability in areas that for the most part lacked a stable farm population. Families living in these communities can often trace back their ancestry for several generations, and farmland has remained in the hands of the same family for a hundred years or more. Usually a combination of strong group identification, religious solidarity, and dedication to farming molded these communities. In the Central Valley, too, despite the more urban nature of settlement and a sophisticated economy, there are ethnic enclaves that display many of the same characteristics.[8]

Ethnicity played an important part in helping to explain community quality in rural areas. In small midwestern communities, patterns evolved based on attitudes toward the land, farming styles, and inheritance. A "yeoman tradition" stressed commitment to the land and the replication of family ownership over several generations, values that made for strong communities. On the other hand, a "Yankee tradition" emphasized entrepreneurial values and concentrated on short-term economic gains. Yankees possessed little attachment to family land and deemphasized community ambiance and social networks.[9] In essence this is a typology to explain farming behavior, and analagous to that used above but in a different form: the familial versus individualistic explanation for the evolution of rural social structure (table 10). It would seem that ethnicity, when combined with familial solidarity, did have a hand in building a sense of community cohesion, whose most obvious manifestation was the survival of small towns and villages where there was a homogenous population.

Familism was a sound vehicle in which to survive the shocks and deprivations characteristic of agriculture for several decades in the first half of the century in the corn-belt and Central Valley. Having survived, close-knit families were in an enviable position to take advantage of the modernization of agriculture, to employ the methods and espouse the ideology derived from a larger world dominated by corporate institutions serving the agricultural sector. Ironically, only a few decades before, "ethnic" farmers had been derided by "progressive" elements in farming for their Old World, inward-looking ways.[10] Down the years, however, the familial tradition was able to blend its value system, which stressed family and community cohesion, with that of the "new middle class," which stressed education, achievement, and strong belief in the corporate business system represented by agribusiness firms. Since World War II the familial and individualistic traditions historically discernible among farm families have melded. The resulting homogenization of behavior has transformed farming practices and economic and social behavior. Here, for want of a better term, I call the new ethic corporate ideology. It has been absorbed over the years from constant contact with outside influences composed of the media, agribusiness, and institutions such as the Cooperative Extension Service.

Catholic Farm Families

From its settlement in the 1850s onwards, the demographic regime in the Catholic community of Auburn in Fayette County, northeastern Iowa, required families to provide occupational opportunities for their children other than those immediately available in the community. Some had to choose work outside agriculture. In the nineteenth and early twentieth centuries, men tended to move into craft occupations, and by World War II, occupational mobility through education allowed farm boys to achieve white-collar jobs. Nevertheless, farming still attracted the majority of children.

The Gruhn family was determined to keep its children in farming. To accomplish this objective the Gruhns employed classical familial behavior with unmistakable elements of entrepreneurial flair. This strategy enabled five successive generations to maintain themselves in farming.

Great-great-grandfather Gruhn first homesteaded in Auburn in 1853. In 1882 he made a predeath transfer to one of his sons, who did the same when the next generation assumed ownership in 1923. Both these intervivos transfers included maintenance agreements drawn up to ensure that in exchange for land regular annuity payments were paid and the older generation received upkeep and board.[11] From the 1920s until the present we have a comprehensive record of one branch of the family, which illustrates rather well a family strategy in which expansion of the base of operations was given high priority.

The second son, William Gruhn bought ninety acres for nine thousand dollars from his father in 1923. In 1916 he had married Sarah Muehl the eldest daughter of a farmer from a neighboring community. Family account books kept by her reveal that four commercial activities supplied the bulk of William's income; he produced cream, hogs, and eggs and also participated in a cream-hauling contract with the village cooperative creamery. In 1924 the family made over $1,200 on its cream operation alone; hog sales brought in $871; eggs $362; and the hauling contract over $1,000. In December 1924, in a month of fairly high operating expenses, the family spent roughly fifty dollars on grocery goods, together with sixty-five dollars on flour and bran. Other expenses included an automobile license for $2.50, and a doctor's bill for $25.00. In addition, $14.45 was spent on tuberculosis testing for the herd and over $250.00 for tiling fields and purchasing soybeans for cattle feed. Egg sales to the local grocery store provided the family with enough cash flow for everyday needs. Additional bills for heavy labor showed that the children were not yet old enough to participate fully in running the farm.[12]

The Gruhns' income and expenditures from 1923 through 1938 show the sharp downward spiral of the economy after 1928 (table 11). Income at the depth of the Depression, in 1932 and 1933, plunged to half the level of the good years, but through prudent management, the Gruhns were also able to trim expenditures correspondingly. From a detailed breakdown of accounts in 1933, it seems clear that the loss of

Table 11. Gruhn Family Income and Expenditures, 1923-1938

Year	Income	Expenditures
1923	$3,571	$3,646
1924	4,397	3,383
1925	3,790	3,779
1926	4,538	4,387
1927	4,083	4,089
1928	4,293	3,642
1929	3,904	3,454
1930	3,518	2,767
1931	2,846	2,424
1932	1,822	1,746
1933	1,790	1,770
1934	2,531	2,402
1935	4,075	3,923
1936	4,798	5,565*
1937	4,712	5,004
1938	5,784	5,311

Source: Gruhn Farm Account Book.

*Includes repayment of $1,021 note.

the hauling contract caused a major dent in the family's income; earnings were reduced to $564 from hog sales, $484 from cream, and $279 from eggs. In December 1933 expenses were $104.00, which included only $14.00 for consumer items, including $4.28 to Montgomery Ward, and $1.18 for groceries. Expenditures were highest in September that year, when children's illnesses and the birth of another child cost the family thirty-five dollars in doctor's bills. Taxes had to be paid on the land, to the tune of $36.78, and two mortgage payments were also due that month, one to William's father and the other to a neighbor. The fact that the Gruhns were still able to contribute thirty dollars to the church in September and to continue to make payments to their parents, indicated their determination to meet their obligations. At the end of 1933, the account book listed a number of obligations. A note of a thousand dollars, at 5 percent interest was owed to Sarah Gruhn's father, along with five seperate notes outstanding to William's father as well. Intergenerational borrowing was always crucial in any family-oriented farm operation; in the depths of the Depression, it was even more important, for familial connections allowed periods of grace that certainly would not have been granted by financial institutions. Under the circumstances, William made only three payments during a 1933—

one to his father and the others to a neighbor who held a mortgage on the farm. Obviously, the family was paying just enough to provide breathing space until better times returned.[13]

After 1933 farm prices improved, and the hog enterprise began to make money for the family—well over two thousand dollars in 1936, 1937, and 1938. By this time sufficient money was available to pay off some of the notes, creating small deficits in 1936 and 1937. The Gruhns were a large family even by Catholic standards, with eleven sons and three daughters. As was fairly common among Auburn families, at least one child was expected to seek a vocation in the church. In the Gruhn's case, their eldest son, with the encouragement of the local priest, was sent to a preparatory boarding school in Texas to begin training for the priesthood. Although instruction was free, the Gruhns' had to stretch their slender resources to provide all incidental expenses.[14]

The tail end of the Depression might seem an inauspicious time to launch a large number of sons on farm careers. However, in 1938 William began what proved to be a remarkable strategy to set his sons' farm careers in motion. Since no land was available in the immediate vicinity of the home farm, William began seeking land in a community within commuting distance. Some years before, other parishioners, with the encouragement of their priest, had begun buying land around Lawler, an Irish community that had the advantage of already supporting a Catholic church and parochial school. The farms were somewhat worn out, and so the predominantly Irish farm population was willing to sell out to the Germans. Together with other Catholic families with large numbers of children, the Gruhns helped initiate a migration that solved the land shortage in the home neighborhood. The Lawler area proved ideal for this purpose. Because the soil was worn out, land was cheap, and abundant family labor made it possible to reclaim the run-down farms.

From 1938 until the end of the 1950s, the Gruhns launched nine sons in farming in the Lawler area. The process was similar for all of them. William negotiated low-interest loans with the Farmer's Home Administration or Land Bank for run-down property, for these institutions would deal only with an established operator. William would then sell the land to one of his sons on contract. Each received between 160 and 200 acres and some livestock to start him off. The typical transaction between father and son set a price of one hundred dollars per acre—well under market value—with a liberal repayment plan. All the Gruhn sons bought their land at a discount price regardless of its increasing cost on the open market. Although the expansion program was interrupted somewhat by the war service of several of the sons, by 1946 the family labor force was again ready to pitch in to make the

Lawler farms habitable and productive. Fences were torn down, outbuildings and houses repaired, weeds eliminated. A small lake was even drained. William and his younger sons commuted daily from the home place, and as each son was established, all the others pooled their labor.[15]

By the 1970s the next generation of Gruhns was ready to move into farming. Their parents employed essentially the same strategy to launch their own children on the land. Each bought a new farm from his father at a price below market value, despite the inflationary spiral in the cost of land. In one branch of the family, five sons were launched in this way; in another, three. By the 1980s family members had exhausted the possibilities of the Lawler area and had started migrating north to another community.[16]

While the achievement of the Gruhn family over the past 130 years was somewhat exceptional, hundreds of other farm families acted in similar fashion. The key to their success was to be able to take advantage of whatever opportunities were offered in a world of limited horizons. These came in the form of low-interest loans, the judicious use of commodity programs, a good knowledge of the land market, and the continuous use of family labor to work towards a common goal. Above all, authority rested in the family head, and spouses played an important supportive part behind the scenes—raising the children, keeping the books, and often taking part in management decisions. At a time when the farm economy was in shambles and many families were leaving farming, the Gruhns proved that a familial approach could make a success of a farm career and could coalesce the needs of individual family members to achieve careers for many of them on the land. At the same time, the determination to find new sources of land was decidedly entrepreneurial. In the past twenty years its involvement with the local bank and its emphasis on encouraging younger family members to go to college before starting to farm, indicated that the younger generation, though still utilizing familism, had absorbed the corporate ethic.

In Auburn, the Depression and its aftermath created a kind of frontier of opportunity that enabled many children who might not ordinarily have had the chance to begin farm careers with some prospect of becoming established. The reinvigoration of farming as a result of the war paralleled a number of other developments that affected occupational opportunities for farm children. Farm boys could be deferred from military service, for example, and depending on the situation at home, some kept themselves out of the army. Many others, however, volunteered for the draft. War permanently altered their lives and their plans, and many never returned to agriculture. A second change occurred in the improvement of educational opportunities at

Table 12. The Children of Carl Wenger

Child	Spouse	Occupation
Albert	Sylvia Holtaus	Farmed, southeast of Auburn
Helen	Florian Nieuhaus	Farmed, southwest of Festina
Ray	Alvina Schmitt	Farmed, northeast of Waucoma
Irvin	Melita Steinlage	Farmed, southeast of Auburn
Irene	Nilus Holtaus	Farmed, north of Decorah
Alma	Richard Koch	Lumber business, South Dakota
Frieda	Linus Mihm	Farmed, east of Auburn
Richard	Geraldine Sloan	Salesman, Osage
Leo	Arlene Reicks	Farmed, homeplace
Irma	John Preston	Insurance, Texas

Source: Kuennen, ed., *Wenger Genealogy.*

the secondary level. Farm children who showed promise in academic subjects were able to seek opportunities in government service or the burgeoning new agriculturally related industries and services. These alternative occupational possibilities eased pressures on the land.

Since the 1940s, even the most farm-oriented families have become aware of dwindling opportunities on the land. Despite its isolation, Auburn showed a surprising awareness of the need for educational attainment to achieve occupational mobility. From the late nineteenth century onwards, men and women chose vocations in the church, reflecting in part the early acceptance of academic achievement and a favorable attitude to higher education. Thus the proverbial bias of farm families against the education of their children beyond the eighth grade was less pronounced in Auburn. One Auburn family well illustrates the changeover from a totally agrarian base of operations to one where some children stayed on the farm but others routinely took up professional or business careers.

Henry Wenger settled in Auburn in 1864, married a neighbor's daughter, and raised nine children. Some of them remained in the parish and either farmed or married farmers. Carl, the eldest, married a local girl and farmed a place he bought himself near the village until he retired in 1951. Seven of Carl's ten children remained in farming. As table 12 shows, two of the girls married men in nonfarm occupations, and one son eventually became a salesman. However, the remainder of the children settled on farms, if not in the home township itself, then quite close by. Carl transferred some property to Irwin in 1946 and some to Robert in 1950. The home place went to Leo in 1960. The importance of birth order and the timing of the start of a career is

Table 13. The Grandchildren of Carl Wenger

Children of	Occupations
Albert (farm)	2 farm, 4 white collar, 1 blue collar
Helen (farm)	1 farm, 3 white collar, 1 blue collar
Ray (farm)	1 farm, 5 white collar
Irvin (farm)	3 farm, 3 white collar, 2 student
Irene (farm)	2 farm, 7 white collar, 3 student
Alma (farm)	5 white collar
Frieda (farm)	2 farm, 3 blue collar, 2 students
Richard (nonfarm)	All nonfarm
Leo (farm)	2 farm, 2 white collar, 1 blue collar
Irma (nonfarm)	All students

Source: Kuennen, ed., *Wenger Genealogy.*

apparent in the Wenger family in the case of Clem, Carl's fifth brother. None of Clem's ten children entered farming. As they came of age, their options were broader for work outside agriculture, while opportunities within farming were more restricted.[17]

Carl's grandchildren, who began their careers in the late fifties, also used educational opportunities to launch alternative careers. Many of them achieved high levels of occupational mobility outside farming (table 13). Some became doctors, lawyers, engineers, bankers, and teachers—exemplars of how farm children could achieve in the white-collar world. But a core remained on the land. Carl's second son, Leo, followed the traditional pattern, dovetailing the beginning of his farm career with the end of his father's, and taking over the home place as a going concern. Later his two younger sons joined him in partnership. In the seventies the family made a conscious effort to expand the farm to use their son's labor and talent to the full. They redesigned the hog and dairy enterprises not only to support three families but also to keep the family members occupied. Meanwhile, other members of the family pursued careers elsewhere. One son became a doctor in a large midwestern metropolitan community.[18]

Obviously, not every family in Auburn was so successful in its strategy for dealing with declining opportunities in farming. Indeed, a fair number of children worked in packing plants or at other blue-collar occupations. However, even in those families with a less than striking record of achievement off the farm, virtually all were able to keep the home place going and to keep it in the family, attesting to the strength of family cohesion and contributing to community solidarity.

Long-Distance Migration

Strikingly different from the short-distance move by German Catholics in northeastern Iowa, was the long-distance migration of Illinois farmers to Iowa in the fifties. In one sense this migration process resembled that of the modern corporate executive, but at the same time, it was reminiscent of the pioneering that took place in the nineteenth century. Although the movers were usually assisted by parents or relatives during the move and land was often bought for them, once in Iowa they were cut off from kinship ties and had to fend for themselves. Family interaction, which most farm families heavily relied on in almost every phase of their social and economic lives, was not possible. In economic terms it made sense to move to Iowa, where land was a bargain compared to Illinois, but in many cases the migrants were also motivated by a desire to escape to an entirely rural environment, where rural values were still unimpaired by urban sprawl. For some families, the migration solved problems and fulfilled certain desires at the same time. It enabled a young family to start farming immediately on cheap land, releasing them from their home areas where pressure for land had stifled opportunities.

Such was the situation for the Paulsens, who migrated to a farm east of Oelwein in the late fifties. The move to Iowa allowed a younger son and his bride to begin farming after the premature death of his father. The father had purchased the land in Iowa so that an elder brother could farm there. After the father's sudden death, however, the heirs agreed that an advantageous method of smoothing out complex estate settlement would be to have the younger son move to Iowa.

The challenges proved formidable, for in a situation reminiscent of an earlier pioneer era, the farmstead was almost unlivable and unworkable. The land, though it had considerable potential, was so wet that only parts of fields could be planted with any hope of producing a crop. Accordingly, Paulsen began a major effort to tile out the farm completely and to build a comfortable home. Over the next fifteen years, the Paulsens reared a family, integrated themselves into the community, and began to expand by buying farms themselves to add to the home place, which was still rented from his father's estate. The Illinoisans, who brought some new ideas with them, stimulated the diffusion of modern techniques and machinery in grain farming and quickened the conversion from small, diversified farms to medium-sized enterprises specializing in hogs and grain. This innovativeness repeated itself in the seventies, when the Paulsens joined with a number of neighbors to form a Sub–Chapter S hog farrowing unit. A few years later, after receiving a substantial inheritance from a relative in Illinois, they formed their own closely held farm corporation and built a farrow-to-finish hog facility of their own, to make themselves self-sufficient and to pave the way for their two sons to enter the business. For the Paulsens the move to Iowa had many benefits, but it hardly represented familism. In fact, the original move to Iowa hinted at a desire to avoid involvement in family affairs. Nevertheless, the move certainly reaffirmed their commitment to farming and rural life. Moreover, it brought them to an area that could benefit from their leadership. The Illinoisans helped transform this part of Iowa into a more productive and progressive community.[19]

All told, roughly forty Illinois families arrived in the area between the middle fifties and the early seventies. Despite the insularity of rural society, most made the transition successfuly. Only a handful moved back to Illinois. Several remarked on the friendly reception afforded them by neighbors, though some old timers were inclined to aim jibes at "the newcomers." One became a county supervisor and another a prominent agribusinessman, who pioneered the transition of grain transportation from rail to road. The Illinois families not only presided over the transformation of agriculture in the area, they also oversaw physical changes in the countryside. A good many of them were from the Grand Prairie area of Illinois, where the tiling of wet ground had originated. These skills were fully utilized in northeastern Iowa as well and transformed hopelessly unproductive ground into land as rich as any in the corn belt. Though never numerous enough to form a colony—in any case they were too individualistic to be clannish—the Illinoisians left their mark on their adopted communities. Even today it is possible to discern a difference between an Illinoisan's homestead and that of a native-born Iowan, for the former invariably has a "showplace" look about it, with modern, freshly painted farm buildings and immaculately landscaped surroundings. This migration succeeded in its long term aim of perpetuating their families in agriculture, enabling their children to make successful careers on the farm. As important was the eagerness of several families to take advantage of the opportunities available in their new environment. They, more than the native Iowans, became leaders in the new corporate-style agriculture of the sixties and seventies.[20]

Protestant Family Strategies

In the German Protestant community of Kane, in Benton County, Iowa, lower fertility among farm couples reduced competition for land. And even if families were above average in size, farmers in the community prided themselves on their ability to leave their children a landed inheritance. As one inhabitant neatly summarized, "The quality of life is measured here by how much you leave to your children."[21] There were about twenty core farm families who could trace their roots back to pioneering ancestors who had first settled Kane in the late 1860s and early 1870s. Among the Germans, most were Holsteiners, who came from a part of Germany where there was a landownership tradition and where independence was prized.

Often in the nineteenth century, immigrants who wanted to farm, spent some time in an urban area before learning of the opportunities available in the countryside. This was the pattern of Isaac Kettering, who, with his stepbrothers, came out from Chicago in the early 1870s. The Chicago fire had destroyed their place of work, but with the contacts they had made there they were able to secure financing to begin farming the virgin prairie. An Englishman from Yorkshire, he married a Holsteiner, and by the time his two sons had begun farming after the turn of the century, the family had more or less integrated itself into the community. In 1905 Isaac retired, and set his sons up on adjoining two-hundred-acre tracts. William, the eldest, had two children; the second son, George, had six boys and two girls.[22]

George Kettering inherited the German work ethic from his mother and some English eccentricity from his father. He was one of the first operators in the community to go into the cattle-feeding business, buying cattle from the West, fattening them on Iowa corn, and then shipping them to Chicago for slaughter. George was a "character" and a superb judge of cattle and the cattle market—a risky business at the best of times. His lack of formal education did not prevent him from being successful, even during the Depression. As one of his principal shareholders of a local bank, he intervened in 1933 to keep it from going under, and some of his shrewd land deals were no doubt assisted by his banking connections. George was one of the old school who never borrowed money to pay for land. He created a sensation in 1927 when he paid seventy thousand dollars cash for a three-hundred-acre farm.[23]

Like many of his generation, George believed that his children should work hard to earn their inheritance. His style was that of the traditional German patriarch, who worked his children and demanded their obedience, all the while using their labor to accumulate farms in the immediate neighborhood. Before World War II, George bought three farms in this way; another followed in 1942, and then two more, in 1948 and 1950. There was never any question as to who was in charge. Indeed, on one occasion a son ran away because he was unable to bear his father's hard-driving behavior. George demanded that his sons work for the family when they were young, promising each that he would eventually receive a farm as a gift. The sons later joked sardonically that their mother had died relatively young in 1939 because of overwork; in any event her death made their father even more eccentric—his unkempt appearance was a legend in the community.[24]

By the early 1950s all the Kettering sons had taken over farms bought by their father, and both daughters had married local farmers. These transfers had taken place in a period when both inheritance and gift taxes were minimal, and the intervivos transfer was the best method of handover. George Kettering's strategy was typical of a period when hard work, frugality, and shrewd business sense allowed a family to purchase land without indebtedness and so bring children into farming.[25]

While the third generation of Ketterings had to contend with their fathers' whims, they matured at a period when farming was profitable. For example, George, Jr., the youngest of the six sons, who took over the home place in 1950, had a relationship with his three farming sons markedly different from his relationship with George, Sr. He and his sons enjoyed a relaxed relationship, and their farm partnership was harmonious. All decision making was done around the table, with all three sons and their wives present. After the passage of the 1976 inheritance tax regulations, George, Jr., sought out an experienced tax lawyer for expert advice on estate planning. At that time, because of use-valuation provisions, it seemed advantageous for George, Jr., to retain ownership of all land for the immediate future. In addition, all family members agreed that an investment in the land market was a wise decision. Accordingly $250,000 was borrowed from a life insurance company to purchase additional land.[26] The choice of a professional from a distant city to advise on estate planning and the participation of all family members in decision making illustrate the changes in family farming. While familism remained an important ingredient in the way they operated, outside influences guided farmers to different behavior. Operating style changed from rigid paternalism to egalitarian cooperation.

In the sixties and seventies, as horizons widened and the corporate ideology gained greater acceptance, farmers absorbed these mores and increasingly watched the bottom line. Work habits and attitudes towards leisure changed. "In my part of the country," one Iowa farmer noted during the high tide of inflation, "nobody raises cattle any more. They all want to play golf on Thursday at the country club, and rub elbows with the doctors and lawyers. They can't do that and raise stock. We've got three main crops here...corn, beans, and Florida."[27] Such changes, can also be attributed to the changing structure of agriculture. In Kane the profitability of cattle feeding was given a severe blow on small and medium-sized operations not by a decline in the work ethic but rather by western competition. Although hogs had taken the place of cattle to a certain extent, grain farming, especially the raising of seed corn, was the usual substitute. Grain farming—at best a part-time occupation—left some Kane farmers brought up with a strong belief in the work ethic with a good deal of time on their hands.

Besides golf and winter vacations, another, more dangerous innovation during the period of inflation, was the new interest in hedging and speculating on the commodities market. Even for those who had the resources and technical knowledge to play the markets, speculating incurred considerable risk.[28] Some farmers would find their initiation expensive. It could be argued that farming has always been a gamble and that playing the markets carries no greater risk than planting a crop in the spring, but being burnt on the commodities market usually turned out to be a humiliating experience. Certainly this was the case for Johnnie Gruber, who, though he managed to land on his own feet, destroyed his relationship with his son in the process.

Before his personal setback, Johnnie Gruber had a very good understanding of his position, that of the community, and where his own children fitted into the scheme of things. He realized that, because of the inflation in land prices and the shortage of land, farming had limitations as a career, and therefore he encouraged his sons to go to college. At the same time, Johnnie also took very seriously the responsibility of providing his heirs with a suitable patrimony. Since he could not provide them with land, as his father and grandfather had done for their sons, he concentrated on life insurance, hoping to give them the kind of liquid assets farmland could not guarantee. Johnnie himself told the story that his father preferred the poker table to the field and described how as a boy he did all the chores while his father played cards at the feed store. In the Depression his father was reputed to have kept the farm alive with his poker winnings. At any rate, when one of his sons returned to the farm from a white-collar job in the city, Johnnie let him take over the operation. Johnnie took an early retirement from farming and began to devote more and more time to the markets.

In the early eighties, most of his working day was spent planning strategy. The tools of his trade were the telephone, a subscription to the *Wall Street Journal,* and a pocket radio. At least three days a week he went to a local sale barn, not so much to buy or sell livestock—although he would buy an animal or two on short-term speculation—as to pass the time of day in congenial surroundings. His real interest was what happened not in the sale barn but in the Board of Trade in Chicago. On the hour and the half hour, the little pocket radio was produced and the market report checked. If an adjustment was needed, Johnnie would move to one of the telephone booths to call his broker.[29]

His world collapsed when, like many others, he overextended himself. He panicked and, seeking to clear up his debts immediately, sold most of his land, including his son's standing crop. His son, who was not consulted, lost half his livelihood with the sale.[30] Ironically, despite the alienation of one of his children and his sisters, who had an interest in the land as well, Johnnie seemed to recover a measure of economic equilibrium, if not his pride. By a quirk of fate, he had sold his ground before prices began to fall. At least he was able to console himself that he did not lose his land like his neighbors—sitting by and helplessly watching deflation eat away at his equity.[31]

Before his fall, Johnnie epitomized a kind of macho style of partying, risk taking, and pride in materialism that characterized farmers in the era of inflation. His use of insurance policies as a strategy for estate planning and his keen interest in the markets were two of the latest trends. Both showed a concern for the well-being of his family. Unfortunately, when action was required that called for a good deal of sophistication and concern for the interests of others, he appeared to forget his obligations and resorted to conduct more befitting a peasant patriarch in the old country than a polished agribusinessman.

Central Valley Family Strategies

California, as the birthplace of agribusiness, whose hallmark has been modern, efficient operations, might be thought to lack the familial pattern in farming. Yet, as has already been shown, California received more than its share of Old World immigrants in farming, and some of these groups historically employed familial behavior patterns. By the late thirties and early forties farmers with a tradition of familial solidarity had adapted to conditions in the Central Valley rather better than old-stock farmers reared in the individualistic tradition.[32] It was their heirs who benefited from the enormous thrust of modernization that characterized Central Valley agriculture after World War II.

California, and the Central Valley in particular, witnessed a process that has already been alluded to in the Midwest: the marriage of conservative, often ethnic farm families with corporate agribusiness service sector—or in other words, a blending of the familial ethic with the agribusiness style. Many of the innovations in marketing, finance, and customs farming had, of course, been developed years before, and so, when a stable, hardworking, frugal, and tenacious group of ethnic farmers established themselves, they were able to embrace many of the trappings of corporate agriculture. Thus, it can be argued that California agriculture, always the innovator, pioneered and developed a new-style operation based upon the highly capitalized family-managed farm. As agriculture modernized in the fifties and sixties, this style of operation gradually diffused throughout the rest of the country.

The desire for a family to succeed in agriculture was probably even more powerful in the Central Valley than in Iowa. By any criteria families who remained in full-time farming in Kings County were successful, and their success was marked by perseverence through some trying times. Even some of the large lake bottom farmers originally started with little, and most families have pursued a quest to endure. In the early history of the county, absentee ownership and outside investment played a significant role, and it was not until the Depression and the bankruptcy of many of these landowners that locally based operators were given the opportunity to expand. These Central Valley families were able to remain in farming by practicing a number of strategies, including the purchase and expansion of a land base; the utilization of family members in the business; the exploitation of resources, such as water, to their advantage; and a flexible approach to the problems of dairying and growing crops under highly competitive conditions.[33] In the boom of the late seventies, their standard of living reached levels comparable to any in the country. Second homes, swimming pools, planes, horses, and overseas trips gave evidence of the rewards that their hard work and foresight had brought.

There was considerable differentiation among the farmers of Kings County—between the lake bottom farmers and those near Hanford, between those growing cotton and those producing fruits and nuts, between family-type farms, larger-than-family farms, and the industrial-type ranch operation. The huge scale of operations on the lake bottom and on the west side ensured that the terms "big" and "small" farmer had some validity. However, the diversity of backgrounds in the Kings farm population belies uniformity in their experience. Most expanded landownership, practiced intergenerational cooperation in management and estate planning, and were deeply involved in and committed to their communities. Even the lake bottom farmers behaved much like the hard-driving midwestern patriarch, but on a much grander scale. Indeed, the central characteristic of several of these families was land accumulation, regardless of the cost to neighbors and business associates.

The Portuguese

The transformation of the Azorian Portuguese from dispossessed peasants to affluent, landowning dairy farmers was a remarkable achievement. The homeland of these farmers, the volcanic Azores in the Atlantic, provided a meagre living in the nineteenth century. The landlord class on the islands were all mainland Portuguese, and the leaseholds on tiny farms were passed down from father to son. For generations islanders had furnished crew members for the Atlantic fishing fleets, and the Portuguese colonies in New England were founded by islanders involved in the fishing industry. The settlement of Portuguese in California came somewhat later, when some whalers landed at West Coast ports. Others employed chain migration via New England to come to California. By the 1880s there were a number of Portuguese working as sheep herders on the west side, and it was they who realized both the potential of dairying in the Kings delta and the utility of employing family labor. To learn the business, they milked for other farmers while accumulating enough money to rent a few acres and buy some cows. Thus the Portuguese transcended the blockage in the tenure system that so many Californians found difficult to breach and found a way to move from the status of laborer to that of operator. Stories of Portuguese ascetism and deprivation in the early days are legendary. Until after World War I few families had managed to attain a secure existence, and the thirties were an especially trying period.[34] Large families toiled on ranches as small as twenty acres to produce fruit, milk, and much of their own food as well. Most farms had no modern conveniences, and often the only water available was from nearby ditches. Typhoid was often a problem. Diets were meager, with main meals consisting chiefly of eggs and beans. Tenant families moved frequently from one farm to another, supplementing their income by hiring themselves out for fruit picking, hay baling, and cleaning canals and levees. However, because the Portuguese quickly gained a reputation for reliability and hard work, they were often sought after as renters. During the Depression, when rent payments could not be made, landowners were sometimes sympathetic, knowing that once the economy improved the Portuguese would merit the risk and consideration. Interviews suggest that, while the Portuguese were welcomed as tenants and dairymen, they suffered discrimination by lending institutions, in education, in the workplace in town, and in voluntary associations. It was only after 1945 that their economic strength forced the rest of the community to open closed doors.[35]

The Garcias were typical of the Azorians who have used migration to better themselves in America. The first member of the family arrived from the island of Pico in 1905 to settle in Wasco, Kern County. His brother joined him in 1911 and then moved to the west side, where he herded sheep for eight years. After the war, he married a Portuguese girl from the San Francisco Bay area and bought a forty-acre ranch in northern Kings County. There, they began raising a family, grew vegetables, worked a thirty-cow dairy herd, and sold butter and cream to the local cooperative. The ranch relied on family labor entirely, and the children had a rigorous year-round regimen of chores. When it came time for retirement in 1942, one son had already entered the navy, so Manuel Garcia stayed on the farm to run the dairy until his brother returned to assume a full partnership. Between them, they bought more land through the Land Bank and gradually shifted towards viticulture. When they eventually split up in 1963, the brother took the dairy and Manuel concentrated entirely on grape production and row-crop farming. By the early seventies, a third generation of the family was ready to enter the business. To accommodate them, the family launched an expansion drive. The purchase of 470 acres with mortgages from Equitable Life came before land had risen above two thousand dollars an acre.[36]

By the eighties the family farmed twelve hundred acres. About half was planted in cotton and corn silage, and the remainder in grapes. Father and son shared management responsibilities, with the row crops assigned to the son and the grapes to the older generation. Seven varieties of grapes were planted for raisins and wine, and the growing year was so organized that harvesting was scheduled as each variety ripened in turn. Harvest began in August with Fiesta raisins, followed in early September by Thompson seedless raisins. Small acreages of wine grapes were harvested in late September and October. The operation employed six men year-round as irrigators, tractor drivers, and pruners. During harvest, as many as forty men were hired individually, most of them coming year after year from Mexico. Rather than use a contractor, the son, who spoke Spanish fluently, acted as crew boss.[37]

The Garcias, then, illustrate how the Portuguese transformed their lives. In many families children whose parents had kept them out of high school in the thirties because they were needed on the farm sent their own children to college. In addition, unlike many who, having achieved a modicum of success, forgot their origins, the Garcias were sensitive to their responsibilities as employers of immigrant labor and to the need to foster good relations with their employees. Though the structural configuration of their operation is a larger-than-family farm and virtually all the work is now done by Mexican employees, they have never used contractors, preferring to remain in very close contact with their labor, whether it is permanent or temporary.

Although some Portuguese families diversified into other areas after World War II, dairying has always been the mainstay of Portuguese farming in the Central Valley. In the Kings-Tulare area these dairies are not the vast factories of southern California and Arizona, though they are large by the standards of the Midwest. About average is a 350-cow herd, employing half a dozen milkers. By definition, therefore, virtually all are larger-than-family-type farms.

The secret of Portuguese success in the industry is the maximization of resources at the lowest possible cost. Seventy years ago large families performed the backbreaking work on an unautomated farm. However, while large families and extremely paternalitic behavior were probably the best strategy for a struggling peasantry to use to establish themselves in California, such behavior became redundant several decades ago. Unless a family lowered its fertility and adopted a more flexible attitude to intergenerational relations, the business was bound to fail. Either there would not be enough land to accommodate all the children, or they would have become alienated by the rigorous regimen required of dairying and quit the farm.

Small families and good parent-child relations, were essentially the strategy adopted by the Cardozas, whose grandfather came from the Azores just before World War I. He spent a brief period as a shepherd before becoming a milker on a ranch near Hanford. He then moved closer to his brother at Tulare and began renting a dairy in 1925. Manny Cardoza worked for his father without wages until he was twenty-one, when he was given thirty acres and thirty cows. In 1944 he moved back to Kings, where he rented 160 acres from a prominent Anglo rancher. During these years he ran a sixty-cow herd completely by himself. By 1948 he had bought the 160 acres from his landlord's estate, but it was not until 1955, when he had suffered a nervous breakdown from overwork, that he hired his first milkers to cut down on stress. When his only son joined him full-time in 1962, they formed a partnership, which evolved a few years later into a closely held corporation. In the sixties the milking herd gradually expanded until six men were milking over four hundred cows twice every twenty-four hours. Unlike the many dairies in the Central Valley that marketed their milk through a cooperative, the Cardozas contracted with a private corporation in a nearby town. All their milk went towards the manufature of cottage cheese and mozzarella used in pizza. They concentrated on updating the physical plant of the dairy rather than buying land. On their three hundred acres, they grew only cotton, preferring to buy their alfalfa in order to achieve consistency. Manny Cardoza, now almost seventy, once tried to retire to his house on the ocean, but boredom quickly brought him back to supervise the milking operation. His son concentrated on the breeding program, as well as business and financial affairs.[38]

The hard work and careful management of the Cardozas, coupled with their excellent working relationship, illustrate that in farming, like any other profession, there is no substitute to enthusiasm for the job at hand. Unlike many involved in the government program, they were not forced to cut back production, and the high production of their herd allowed them to stay self-financed. At a time when many dairy farmers have extremely high debt loads, the Cardozas were not vulnerable to high interest rates or to the beck and call of a lender. The family retained many of the characteristics of the "frugal farmer," so dear to lenders in a period of economic difficulty. At the same time, they have adopted the most modern techniques for breeding and for herd and personnel management. The latter is particularly important in California dairying because of the competition for good milkers and the relative ease of employment outside agriculture. Although the job is not physically punishing, its routine can be trying. On the Cardoza farm the milkers, who were all Azorians, were furnished a free house, utilities paid, next to the dairy and the homes of their employers; they were also given a day off every week, retirement and medical coverage, and free milk and meat. These strategies, which combine some well-tried behavior learned many years ago with innovations discussed in the pages of the glossiest farming journals, have achieved their aims. Soon the eldest generation will step aside to leave room for the next to take over. When Manny, Jr.'s only son returns from college, the cycle of management will begin again.[39]

The Dutch

Like the Portuguese, the Dutch have made a sizable contribution to agriculture in the Kings delta area. They too arrived in the Central Valley around the turn of the century. Some worked on dairy farms as milkers before moving to Los Angeles to open drylot dairies of their own. Others stayed near Hanford and began raising fruit. Because of their northern European origins, the Dutch found it easier to establish themselves than had the Portuguese. In addition, the Dutch were a less homogeneous group; most were Protestants, but smaller numbers of Catholics also settled the area, and they tended to be more outgoing in their social relations. Third-generation Dutch Catholics have intermarried with the Portuguese, whereas Dutch Calvinists, with a larger pool of eligible marriage partners, especially from the midwestern Dutch communities, have not had to look outside their immediate ethnic or religious group. The contemporary Dutch presence has recently been augumented by the migration of dairy farmers from Los Angeles. Forced out of business by urban sprawl, most moved north with handsome sums paid out by developers. In Kings they have been able to purchase cheaper ground, which they often had to rehabilitate because of alkali damage.

The oldest and best-established members of the Dutch community have concentrated on fruit growing and general farming over the past seventy years. In the early part of the century they produced fruit for the Hanford packing plants; today, the emphasis is placed on more profitable high-quality, fresh fruit shipped to the markets of the East and Midwest. Obviously, where the objective is to market the highest-quality fruit as early as possible for the highest possible price, days or even hours matter. Crop failures in others parts of the country or weather related problems in the Central Valley affect prices from one week to the next. Soft fruit farming is one of the riskier enterprises in agriculture.

The DeRaad family has surmounted these risks over three generations, farming near the Kings River since just after the turn of the century. Joe DeRaad emigrated from Holland in 1906 and bought sixty acres with his brother in 1910. Their methods were cautious and conservative. Having just managed to hang on during the Depression, they shunned deficit financing, employed Dutch hired hands, and followed life-styles that closely resembled those of the Midwest. Only in the fifties did the family finally give up dairying to concentrate entirely on row crops and fruit. In the sixties a new generation began to make itself felt, and the DeRaad business rapidly developed into one of the most progressive medium-sized farms in the area. Joe's grandson John shunned the yeoman image and embarked on an expansion program that entailed buying and renting land. Eventually the family managed twenty-two hundred acres, of which eight hundred were devoted to soft fruits, four hundred to walnuts, and a thousand to row crops.[40] The operation successfully experimented with interplanting cotton and young walnut trees, financed a modern walnut huller, and negotiated a private contract with some Texas restaurants for a large percentage of the walnut crop in order to bypass unprofitable established marketing arrangements. In the past few years the family corporation was joined by another family member, a younger brother, whose function was to manage some of the more technical aspects of the business, such as fruit spraying and irrigating. Employee relations were also given special attention on this ranch. Every effort has been made not only to give the permanent employees a year-round schedule of work but also to provide the Mexican picking crew a continuous schedule in the orchards. In the fresh fruit business there is no substitute for an efficient labor force.[41] In the recent economic downturn, the DeRaads have worked to reduce expenses in daily operations outside of labor, to mechanize pruning, and to negotiate the best possible price for cotton and soft fruits. Even in a time of austerity, there is a sense of excitement and innovativeness about every aspect of the DeRaad business, an example of entrepreneurship at its best.

The Midwesterner Transplanted

The origins of those who settled Kings County were diverse, but at least a portion of the pioneers came from the Midwest, especially from states west of the Mississippi, such as Iowa, Nebraska, and Kansas.

These families brought with them a Yankee spirit of involvement in civic affairs, and a belief in the importance of political egalitarianism and in a fair division of the abundance agriculture provided in the Central Valley—notions some California ranchers considered unimportant. At least some of the members of these families had come of age in the Depression, and had this spirit reinforced by the New Deal.

The Warners, for instance, came from Mitchell, South Dakota, in 1898 when their hogs were wiped out by cholera. With the help of a mortgage from a Hanford bank, the grandfather and his brother purchased 480 unimproved acres outside of town and tried to replicate midwestern agriculture in the Central Valley. They concentrated on hogs and dairy during the early years until they had absorbed enough information about their new environment to plant Thompson seedless grapes. Farming did not remain the center of their lives. Like many old stock families, the Warners were interested in giving their children as much educational opportunity as possible to enable them to pursue careers outside agriculture if they chose. The desire for education meant a family move from the ranch to Hanford, and the family became better known in the community. In the late twenties one family member was elected a county supervisor, and he continued to represent the small farmers along the Kings River until the fifties. The tradition of service continued in another generation of the family when John Warner's wife, who had a degree from the University of California at Davis, was hired by the Extension Service to conduct 4-H programs for children, and nutrition clinics for mothers in labor camps. This was the first program run for farm labor families in the county, and it caused a stir among the ranchers, who did not welcome government intrusion in the lives of their labor.[42]

In the Warner family, intergenerational transfers were reminiscent of patterns in the Midwest. Because of the Depression, John had difficulty starting his career as a sheep farmer. Banks would not finance him, and the Production Credit Association would only lend money if his father signed a note—something his belief in individualism prevented. Eventually, when credit became easier to obtain, after the war, he was able to purchase a 160-acre ranch of his own.[43] The fourth generation of Warners to farm in Kings, began row-crop farming in the seventies. Although John Warner would have liked his son to begin on his own, as he himself had done, the highly inflationary climate of the late seventies made it impossible for a young man to launch his own career. In view of the developing economic downturn, it was fortunate for the family that Warner took a cautious stance. His son wanted to expand, but Warner refused to allow him to build a new house and also vetoed the purchase of expensive land. Instead, he worked out a series of incentives whereby small amounts of property would be transferred provided his son performed adequately as a manager. Ironically, partly because of the times but more because of parental cautiousness and their own bad memories of the Depression, these third-generation Californians had become more conservative than their own parents had been. John Warner was unwilling to lower the cash rental payments he required of his son, for example. In 1985 the young Warners were forced to reappraise their position. Rather than risk a family confrontation, they left farming, and the Warner land was then rented to a neighbor.[44]

A number of Kings County pioneer families had originated in Iowa, and many kept the beliefs about community involvement and egalitarianism that their ancestors had brought from the Heartland. The Huttons migrated from just east of Council Bluffs, and settled on twenty acres of truck farm south of Hanford in 1896. When William Hutton died in 1903 his estate totaled $4,803 and included one bull, eleven calves, eight hogs, one mare, half a share of ditch stock, a bay horse named Ranger, seven stands of bees, and a brindle cow named Gussy. It was an estate as modest as any left in Iowa at that time.[45] A number of other Iowa families moved into the neighborhood, and all practiced diversified farming, with cattle, hogs, and dairy as their main source of livelihood. Gradually the land in the area was improved; the high-water sloughs were filled in, and the salt was leached from the soil. Throughout these years the Huttons lived modestly; when another member of the family died in 1920, his estate was worth only $9,933.[46] The family managed to struggle through the Depression, working their 240 acres without benefit of hired help. The greatest challenge, however, to both the family and their community, came after World War II when the then-head of the family, Peter Hutton, mobilized the whole northeastern part of the county in a quest to save their ground-water rights from the encroachment of outsiders.

Peter ran a commercial hay-cropping business, as well as a dairy and row-crop farm on 240 acres. In 1951 the big lake bottom farmers were suffering a water shortage. In years past they had bought up groundwater rights from farmers in the northern part of the county, with the intention of using pumps to bring the water to the surface and transport it by canal fifteen miles to their farms on the lake bottom. Most of this activity took place next door to the Hutton's farm, and rather than allow the pumps to suck the neighborhood groundwater dry, Peter started to organize to stop the water piracy. Going door-to-door he single-handedly roused the community. He presided over a mass meeting in Hanford, where the farmers raised enough money to hire an attorney for a class-action suit to seek a restraining order against the lake bottom farmers.[47]

Unfortunately Peter never saw the final results of his efforts, for he died of a heart attack a few months later. Despite this severe blow to the family, they closed ranks around their mother, who with her children kept the farm going until the next generation was old enough to take over. When Joan Hutton died in 1976, she still retained ownership of some parcels of the family land, and her estate was worth approximately $350,000. The family continued its involvement in community affairs; members regularly served on the boards of gins, water districts, and schools in the area. This commitment to service extended well beyond the immediate community in the case of one of Peter's grandchildren. He served for a time in a high-level position in state government before coming back to the family farm.[48]

The Legacy of the Dust Bowl

Kings County was never a primary receiving point for migrants from Oklahoma and Arkansas during the Great Depression, but compared to the population as a whole, quite large numbers did settle in the county between 1935 and 1940. Most remained farm laborers or blue-collar workers all their lives, but some managed to cross the divide between farm laborer and farm owner. Tom Slayton hailed from a sharecropper background in Arkansas. As a young man he moved to Oklahoma in the twenties and sharecropped cotton as well as working in the oil fields. Ambitious and a hard workers, he apparently prospered, making good money even in the depths of the Depression. Like thousands of others, he was lured west in 1935 by tales of opportunity in the Central Valley, but his trip hardly resembled the migration of the Joad family in *The Grapes of Wrath.* Unlike John Steinbeck's characters, he traveled in a car that was practically new. Apart from sleeping on the ground and eating too many bologna sandwiches, Slayton experienced few of the trials depicted in the novel.

Slayton worked in Hanford for five years as boss of a cotton-chopping crew composed entirely of Okies and Arkies. Eventually he sent for his family, and they shared a run-down ranch house with a number of others. Gradually, the family began to accumulate enough money to purchase some land, and by the early forties Tom had the forty-four-hundred-dollar down payment necessary for sixty four acres south of Hanford. From then on, the Slaytons never looked back. Over the years Tom's entrepreneurial flair accumulated over five thousand acres. His four sons all worked in partnership, and by any standard of comparison, the family developed one of the most respected medium-sized cotton operations in the community. In the seventies, the old man and his sons decided to split up the partnership, and the family land was distributed amongst them.

This rags-to-riches story should end there. Unfortunately, Tom's success was marred at the end of his life by a bizarre family tragedy, in which a grandson was involved in the contract murder of his parents for their inheritance. That event belongs to a later chapter.[49]

The Barons of the Lake Bottom

Industrialized corporate agriculture is exemplified by farming on the lake bottom and the west side. Indeed, since the war the present occupiers of land around Corcoran have all farmed on a grand scale. Many of these firms began decades ago as modest-sized operations. The founders were bound by an ethic that combined a desire to expand landownership, a dedication to the profession of ranching, a sense of the importance of family, and a commitment to community in an often-harsh environment. These core values, so useful in building up a family business, were compatible and easily transferable to the corporate style.

Tulare Lake, on whose bottom these families farmed, has no natural outflow. Because heavy clay soils prevent a measurable percolation of flood waters into the ground, water remains on the surface until it evaporates or is used for irrigation. These unique conditions made it impossible for farmers to supply their water needs by pumping groundwater. Water had to be brought in from other sources, and to acquire the necessary water, the operators formed one of the most powerful lobbying groups in California agriculture. They secured direct access to the legislature in Sacramento and to Congress on matters of interest to them. In an area of 188,000 acres, 177 separate owners controlled property in the 1970s, but just 15 of them farmed 90 percent of the land area.

Controversy has swirled around the lake bottom farmers for at least fifty years because of their involvement with water politics and their unyielding stand on union representation for their labor. These ranchers were survivors of the rigors of the economy and the unusual environmental conditions of southern Kings County. They were, according to one observer, "a hardy group that had the pioneer spirit and pioneer opposition to restraint by others."[50] This fierce desire for independence in their own operations often resulted in conflicts both among themselves and with ranchers surrounding Corcoran. As they liked to point out, small farmers were not able to withstand the financial burdens of fighting floods, the temporary abandonment of flooded land, and the special cultural practices dictated by the soil and drainage conditions of the lake. Historically, small farmers either expanded their operations to an enormous size, leased their land to a larger operator, or sold out.[51]

Of the big four—Payne, Wood, Shepard, and Jensen—only the first had not expanded from a family-type operation. Payne was a diversified corporation with headquarters in Los Angeles, too large to be considered a family-style business. The others, however had gradually built farming empires and since the forties, all had been industrial-type operations run by family members. The career of the redoubtable Charles Wood, already described, was more flamboyant, perhaps. Wood used every resource at this disposal to expand landownership and make money in ranching, including the delayed payment of income taxes, the harvesting of other people's crops, and the movement of county property in order to improve the access to his land. The other large landowning families on the lake kept a lower profile and presented a more polished image to the outside world, but all were bent on money making and expansion. Only one of them had immigrant European origins.

Like thousands of other natives of Schleswig-Holstein, Dil Jensen's family was caught up in the exchange of territory between Denmark and Prussia after the Danish-Prussian War. As a teenager he left his home for the United States, migrating to California after short stays in New York and Iowa. At first he worked as a butcher boy in the San Francisco Bay area, but then he made his way to the Central Valley, where he began his farming career herding sheep in the Kettleman Hills. By 1893 he had moved to a fruit ranch near Armona and had married the first of three wives. Land accumulation was the principle motivation of his business life. He bought his first piece of land in Corcoran in 1899, believing, quite rightly, that the Tulare Lake had great potential for development. By World War I, Jensen had sold his fruit business in Armona and moved his operations entirely to Corcoran, forming what was then the largest dairying operation in the county. In the twenties and thirties it was not unusual for Portuguese and Dutch milkers to handle eleven hundred cows at the Jensen dairy. Jensen also assisted in the development of cotton processing in Corcoran, as well as the huge and elaborate irrigation network, which required the digging and dredging of canals to distribute water all over the lake bottom. During the cotton strike of 1933, Jensen was a leading grower organizer, who arranged for Armenian pickers to be brought into the town to break the strike.[52]

With the death of Dil Jensen in 1939, the family divided its operations between his sons from separate marriages. David Jensen's family concentrated on the cattle business, with a feedlot in Corcoran and six thousand acres of grazing land in the Sierra foothills near Porterville.[53]

The other side of the family gave up dairying to concentrate entirely on row-crop farming on the lake. The Jensen operation is now run by Dil's grandsons as a closely held family corporation. Since the forties they have also become involved in the crop-dusting business, which became increasingly important in industrialized agriculture. Their six thousand–odd acres largely escaped the 1983-1984 flood, which inundated much of the Payne and Wood land. And the family has avoided the internal disputes that plagued their neighbors.[54]

Dilemmas of West-Side Agriculture

Farming on the west side has presented still another set of problems to family farmers in the twentieth century. The land on the west side of the lake is of poor quality, and because it lies at the most westerly end of the Kings River delta. In years of below-average rainfall, surface water cannot be guaranteed to farmers for irrigation. However, because there is no confining clay layer below the surface, groundwater can be pumped for irrigation.

The only practical crop in 1933 when the Denton family moved from Corcoran to Stratford was wheat, with barley a possibility when plenty of water was available. The family also owned land west of the Kettleman Hills, which was dry farmed. When Donald Eaton married into the family in 1945, he formed a three-way partnership on a thousand acres with his brother-in-law and his father-in-law. He and his bride bought themselves a share of the partnership by investing twenty-five thousand dollars, followed by fifty thousand dollars the following year. In 1948 the farm did so well with its flax crop that the Eatons were able to pay off their house debt in one year. In the early fifties, with the sinking of deep wells, the operation began on the advice of the Extension Service to grow cotton on the Stratford farm. Labor came chiefly from the Stratford camps, from which blacks were transported out to the farm to chop and pick cotton.[55]

The Denton-Eaton operation had an aura uncharacteristic of row-crop farming. Perhaps because some of their land bordered on the range country of the central coast, they lived and worked more in the style of the livestock rancher whose background and education suggested landed gentry. These characteristics set them apart from families with dirt-farmer backgrounds. While in no sense "gentlemen farmers," the family has greater affinity to large landowners in Australia, for instance, than to their more down-to-earth neighbors.

In 1970 the Eatons' farming operations changed profoundly with the delivery of federal water to the Westlands Water District. On one section of their land and on additional acreage rented from Southern

Pacific, the soil conditions made it possible to grow a whole variety of crops that could not be grown on their poorer ground. Under a crop allotment, they began with cantaloupes and fresh-market tomatoes. However, disease prevention makes it imperative to switch vegetable crops regularly. Thus, the Eatons have contracted to supply carrots, eggplant, tomatoes, and seed lettuce at one time or another to large processors.[56]

By the late seventies the parents were beginning to plan their retirement. Their four sons and one son-in-law had all finished their schooling and wanted more formal positions in the operation. With this in mind, the family purchased a further eight hundred acres near Kettleman City, which, because it could utilize state water, did not come under the acreage limitation laws of the Westlands Water District. In 1984 the family drew up its transfer plans for the closely held corporation formed in 1972. The parents began to transfer shares to their children, giving stock to the sons and son-in-law in compliance with gift tax regulations. They also began to transfer their extensive collection of equipment to the younger generation.[57]

If the Eatons represent something approximating the landed gentry, their near neighbors, the Ingram family, represent the much different tradition of the technocrat. The Ingrams have engaged in agribusiness and corporate farming for two generations. Tony Ingram's father was a civil engineer born in Jackson, Michigan, who worked for a railroad before coming out to California to work for the Kings County Land Company, which in the twenties and thirties, owned large portions of the Tulare Lake basin. He helped build the Tulare Lake canal, which was instrumental in preparing the lake bottom for large-scale farming. He had the chance to buy some lake bottom land in the thirties at $7.50 an acre but turned it down. Tony himself graduated from the University of California at Davis in 1936 and began farming rented ground both in Stratford and at Five Points in Fresno County. After war service, he bought land west of Stratford in 1946 and farmed cotton, alfalfa, and melons with irrigation water from deep wells. In the 1950s, with his wife helping with the business end of the operation, he owned twelve hundred acres and rented another thirty-eight hundred. The fifties were crucial years for west-side farmers. It was then that they realized their sole source of irrigation water, the groundwater beneath their farms, was being seriously depleted. In response to the situation Westlands was formed in 1952. Ten years later construction was begun on the San Luis Dam to provide storage for the water brought down from northern California. But it was not until the early seventies that federal water began to flow as far south as Kings County. Tony Ingram's farming headquarters was located on the Fresno County line near Interstate 5 in the Westlands Water District, of which he is president. His residence is farther east, in the so-called island district of the county. There, he has created a country estate, complete with grazing cattle, a long drive lined by a white fence, and a large home surrounded by trees.[58]

Ingram, president of Westlands since 1977, has had to face two major crises during his term in office. The first concerned the limitation law and the attempt by Congress to enforce the regulations in Westlands, and the second was the uproar created by the selenium controversy. Ingram himself considered the limitation enforcement controversy a greater threat to the livelihoods of his members than the halt of irrigation that resulted from the selenium discoveries. He maintained that the difficulties with selenium, though serious, had been blown out of proportion by media "hysteria"; limitation, which struck at the Achilles' heel of ownership on the west side, was "much more disturbing."[59]

As chairman of the Western Cotton Growers Association, Ingram was also involved in farm policy at the highest levels. These activities and those with Westlands gave him a much more cosmopolitan orientation than the majority of the ranchers in the neighborhood and a corresponding reoriented view of family farming. Ingram's only surviving son has already virtually taken over the business from his father. Apart from some occasional financial advice, the younger generation is fully in charge of the Ingram Land Company.[60] The immaculate shops, the neat headquarters offices, and the apparently contented labor force seem to represent an ideal in larger-than-family-size operations. This is the kind of firm, it would appear, to which the cheerleaders of corporate agriculture would have the family farmer aspire in the next decades.

Conclusion

Since World War II, agriculture in some areas of the Iowa corn belt and in the Kings River delta of the San Joaquin Valley, has increasingly been controlled by family farmers with a record of stability in a particular neighborhood over several generations. Often these families have had immigrant backgrounds, and in Iowa they were able to duplicate inheritance and transfer practices brought with them from Europe. Familism was the best tool for success in the first fifty years of this century. Then structural change in farming demanded different strategies. Family solidarity remained an important ingredient, but long- and short-distance migration, sacrifices for the children's education, and the adoption of new techniques and new crops all played a part in keeping a family business going, whether it was in the corn belt or the Central Valley.

In the new world dominated by corporate institutions, familism dependent on paternalism and fairly rigid hierarchical arrangements had to be transposed into family businesses in which roles were fluid and leadership firm but low-key. As was seen in these case studies, some families were more successful than others in making this transformation work. All kinds of extraneous factors could impinge on a family business and make it fail. Nevertheless, the principal source of failure was the inability of family members to integrate themselves into the firm. The farm business, like other kinds of family businesses, was notoriously vulnerable to disharmony among family members.

7 Community

The hundreds of towns founded in the nineteenth century to serve Iowa farmers formed a continuum from crossroads hamlet to county seat; the main function of such centers was to permit local farmers to conduct their business without traveling too far from home. The same principle held true in the Central Valley. The formation of Kings County in 1893 was a classic case in point. An Iowa-born newspaper editor drummed up enough support for the secession of the area that would become Kings from Tulare County, basing his campaign on the issue that Visalia, the county seat of Tulare, was too far away to allow citizens from the Hanford area to make a round-trip journey and complete their business in a single day.[1]

In his analysis of the Central Valley towns of Arvin and Dinuba, Walter Goldschmidt attributed differences in community quality to the kind of agriculture practiced.[2] As we have seen in the previous chapter, the cultural predilections of the surrounding population also had a bearing on the quality of a given community. In Iowa, certainly, ethnocultural factors must be joined to economic analysis to explain the nature of community. Perhaps a purely economic interpretation holds more validity in the Central Valley, where it was first applied.

In Iowa, small communities of fewer than twenty-five hundred people have typically been in a spiral of decline for many years. As the downturn of the twenties developed into a full-blown depression, many key small-town institutions, such as banks, failed. The full complement of services available in many small communities during the golden age of agriculture were gradually whittled down by hard economic times and the increased accessibility of larger towns and cities made possible by the automobile. Professionals and businessmen either closed up and moved to where better opportunities were available or stuck it out until retirement and then were not replaced. One northeastern Iowa community lost thirty of its seventy businesses between 1907 and 1941. One of the physicians departed, as well as both dentists, and such community gathering places as restaurants and hotels closed their doors. But at the same time, structural changes in the economy brought new services such as automobile dealerships and the telephone and utility companies, which immensely improved the quality of life.[3] A further assault on small towns occurred when the pace of change quickened in the countryside after World War II. Demographic changes among farmers and in land tenure reduced the need for certain services in town, and the elimination of one-room schools and the consolidation of school districts hastened the decline. The school was often the anchor of a community; once it was lost demise became inevitable.

In towns that succeeded in retaining their institutional base, the boom years of the seventies saw a modest revival. Those near larger cities often had some growth as bedroom communities, providing inexpensive housing for residents who commuted to work.[4] However, it was the prosperity of the surrounding farm population itself that spurred the growth of innumerable service establishments, which aimed to capture the farmer's newfound dollars. Many automotive and implement dealers upgraded their plants and services, and retail stores did likewise. Then the optimism the inflationary boom had created on main street was rudely shattered when the farm debt crisis began to have its full effect in the middle eighties. In Iowa, where agriculture was interconnected with every part of the economy, the downturn triggered a statewide recession. Thus the state became the community writ large: as farmers suffered, so did the whole economy.[5]

In the Central Valley, where Goldschmidt's ideas were first tested, the decline of a particular community over time was not as important as the effect of large farms on the quality of life. In his original formulation, Goldschmidt chose Dinuba, an old community by Central Valley standards, settled for the most part by transplanted midwesterners, who farmed small acreages. With it he compared the newer, unincorporated community of Arvin. Unlike Dinuba, Arvin had virtually no surface irrigation to support its agriculture and was also heavily involved in oil production. Ranchers there farmed larger acreages and were engaged in different farming enterprises than in Dinuba.[6] Although it is possible to question Goldschmidt's choices, it seems clear enough that in California there was a correlation between industrialized agriculture and the quality of life in a community. The lake bottom farmers liked to point out that their operations employed the equivalent of three families for every section they farmed, seeming to suggest that their wage workers represented the equivalent of three farm families living on homesteads and fully participating in the life of the community.[7] Unfortunately, they argued in defiance of the facts, as even the most charitable observer of a community like Corcoran would note sooner or later. Essentially, a community dependent on farming for its economic life needed the purchasing power of farm families, and their total involvement in the schools, churches, and voluntary associations of the town enhanced its value as a place to live and work.

By the 1950s the dominance of the California agricultural establishment had been settled. The Central Valley Irrigation Project was built, and industrialized agriculture was in an unassailable position.[6] The leadersip of Central Valley counties such as Kings recognized that an economy entirely dependent on agriculture had limitations. Structural change in farming could only mean larger operations and smaller numbers of farm families to support the local economy. Some attempted to solve this problem by attracting industry or by snaring a military establishment that would provide employment to the local population, bring in large numbers of dependents to boost retail sales and to stimulate the real estate market, while leaving the environment, especially farmland, untouched. The opening of the Lemoore Naval Air station in the late fifties did all these things for Kings County.[9] Later, the state promoted other facilities, such as prisons, on the same premise: that an economy entirely dependent on agriculture was at a disadvantage.[10]

In contrast, Iowa allowed her economy to remain too dependent on agriculture. There, changes in the makeup of the farm population and its enterprises had a much more decisive impact on the economic health of the community as a whole.

The two Iowa centers singled out for scrutiny here are villages where the integration of family and community has occurred over a lengthy period. Retirement of farmers to these trade centers has further strengthened this integration. Strong ethnic and religious homogeneity has fostered a sense of loyalty to the community, and even though some major institutions such as schools have been closed over the years, the villages and their residents have been able to surmount these obstacles. The two California communities are less easy to categorize. Both are larger than the Iowa villages, with more diverse populations and economies. However, farming remains the major force in both, and farm families retain their importance in economic, cultural, and political life.

Hamlet and Village

Auburn, in Fayette county, is a German Catholic parish, which in 1980 had 887 parishioners. Many of them live some distance from their church but return to the community of their ancestors to worship. The small trading center, sheltered by woods at the bottom of a hill and dominated by the twin towers of the parish church, is attractive at any time of the year. When the snow is on the ground or in early summer when the fields to the north show evidence of strip cultivation, the landscape reminds one of Bavaria or Austria rather than the corn belt. The farm yards, flower gardens, and family shrines of two working farms within the incorporated limits of the village enhance the Old World atmosphere. The church, situated on a rise that commands the village, was built by a Dubuque contractor in 1914. Its red brick exterior and high, vaulted ceiling make it a fine example of the Gothic revival architecture of the upper Mississippi Valley. To one side of the church, past the old baseball field, is the cemetery, surrounded by evergreens planted in the last century. The inscriptions on the headstones reveal a good deal about the community. The relatively small number of surnames, in particular, indicates a closely knit, stable parish, which has changed little since its original settlement in the 1850s.

At the bottom of the church hill lies the residential and "business district." The 1910 census lists ten businesses in Auburn, a village of 150 people. Among them were a general store, creamery, restaurant, hardware store, sawmill, and four or five other "merchants."[11] Today the crossroads business district supports a bank, a garage, a hardware store, a supermarket, a jewelry store, a restaurant, and a tavern, along with a post office and a feed mill a little way up the street. Most nights, but especially after the Saturday night church service, the business district is crowded with cars. After church the restaurant caters to the older generation, while younger members of the community patronize the bar next door. The restaurant is one of the more important institutions in the village, for with the jewelry store, it has made a reputation beyond the community and brings in outsiders to spend money in Auburn. Across the street the hardware store and supermarket continue the tradition of small rural retail outlets, stocking a huge variety of goods. The former sells everything from guns and ammunition to microcomputers, toys, frying pans, barbed wire, and bicycles. The supermarket, distinguished by its magnificent polished oak floor, can outmatch the competition in nearby towns with its comprehensive selection and display.

It is a tribute to the resourcefulness of the local businesses and the loyalty of the farmers that a trading center so small can support a dozen or so businesses. However, it is the institutions at the top of the hill, the church and the parochial school, that have dominated the lives of the residents of Auburn. The successful businesses at the bottom are but extensions of over 130 years of adherence to the church and its ancillary institutions. Without the church, there would be no Auburn.

Auburn and its church, like a number of Catholic communities in northeastern Iowa, owes its origins to the missionary activity initiated by Father Samuel Mazzachelli in the 1830s and continued by the first bishop of Dubuque, Mathias Loras. In the late 1840s some German Catholics in southern Indiana were recruited to settle in this part of Iowa. By 1855 the original thirty or so families had built themselves a church, which for the next sixteen years was served by visiting priests from other parishes. Only in 1871, when a new brick church was constructed, did Auburn obtain its first resident pastor.[12]

For years the thick woods and rivers to the east and south cut off social interaction with the old-stock and non-Catholic residents who lived in those directions. More natural alliances were formed with other Catholic communities to the north, northeast, and northwest. Marriages and kinship networks moved in this northerly axis, and Auburn families stayed north of the river until the 1950s, when they finally began to cross into what used to be Yankee territory. At the time of settlement about a third of the land area of the parish was open prairie. The rest was woodland, which had to be cleared and grubbed out before planting could begin. The woods at first contained game, especially deer, but by the turn of the century they had virtually been eliminated. Only in the past twenty years has careful repopulation by the state allowed deer hunting once again. The woods also provided an alternative livelihood to many families when times were hard. From settlement onwards, wood lots were sold in five-acre parcels, so that residents had access not only to fuel but also to lumber to be sold for commercial purposes.

While the thick woods and hilly countryside might have looked attractive, they required a peculiar dedication in order to extract a living. For this reason, Auburn men traditionally took whatever off-farm work was offered them. Carpentry, blacksmithing, and in the twentieth century mechanical work supplemented family income. The soil and topography made dairy farming the most practicable operation, and until it was closed down in the late 1950s, the cooperative creamery was one of the major community institutions. Dairy farming, which has often been described as the modern equivalent of slavery, permits no easing of schedules. Even the introduction of labor-saving devices had not altered the Auburn tradition of round-the-clock hard work.

About eighty miles to the south lies Kane, on the prairie in Benton County. Early photographs reveal a treeless, windswept village consisting of a main street with a dozen or so shops and houses. Indeed, the physical makeup of Kane is in complete contrast to Auburn's. The surrounding country is fairly high rolling prairie, which must have seemed forbidding to the pioneers who settled there in the 1860s, especially in winter. The bleakness of the area is reminiscent of the high plains, but the inhospitable-looking rolling prairie gave the pioneers a good base on which to build a prosperous community. The rich soil guaranteed a handsome return after the sod was broken.

The early history of the community also had a connection with Catholic missionary activity. The fledgling University of Notre Dame purchased two sections of land in 1863 and established a church and seminary, but these foundered after a few years. The actual settlement of the area and the founding of village was by Schleswig-Holsteiners. They came as farmers and village dwellers and would eventually far outnumber the Irish Catholics who also settled the prairie. The Holsteiners brought with them a tradition of political and religious independence. Unlike the pious, hardworking Lutheran peasants who settled the Iowa prairies in the second half of the nineteenth century, they were agnostics. A measure of their disregard for the conventional mores of the times is the delay in establishing a church. In 1895, fully thirty years after the first Holsteiners had pioneered the land and fifteen years after the platting of the village, the first church finally came into being.[13]

Driving through the community today, the casual observer might fail to notice the roadbed of an abandoned railroad line, all that remains of one of the five trunk routes that crossed Iowa. Kane is just one of the many villages that mark the route of the old Milwaukee Road mainline from Chicago to Council Bluffs. The Milwaukee, comparatively late in building its trunk route westwards, bypassed most of Iowa's major cities. The company concentrated, instead, on express passenger service and farm freight. To serve the farmers, the railroad platted over forty small villages at three-mile intervals the length of the state. The village of Kane was one such railroad-sponsored community, with an additional role as a service center for the surrounding farms. Today, because of structural change in farming and elsewhere, church membership, ironically, provides as satisfactory a definition of community in Kane as any other. In 1981 the Lutheran church had 680 communicants.[14]

The abandoned railroad right-of-way, whose track was in place exactly one hundred years before it was torn out, symbolizes the structural changes in the community. Not only has the railroad become redundant as a means of transportation but the major agricultural enterprise of the community, cattle feeding, which the railroad helped establish, has diminished in importance. For years the Milwaukee brought livestock from the West to be fattened on local corn and then shipped out to Chicago. The cattle business, such as it is, is now localized and handled at small country shipping points by truck transportation.

The economic history of the village resembles that of hundreds of other small corn-belt communities. But unlike many others, which have degenerated because of the loss of key institutions and transportation facilities, Kane has moved forward in spite of setbacks. There are almost as many businesses in town as there were in 1910.[15] The affluence of Kane can be measured in its homes. There are no ramshackle, unpainted houses, whose windows are draped with celluloid to keep out the cold. Instead, both in the new addition and in the older part of town, houses stand on their own lots with neat lawns and flower beds. Years of careful nurturing and tree planting have eliminated the raw prairie look. The public park, with its tennis court, picnic facilities, and playground, lies beside the baseball field and the village-sponsored nursing home. As befits the traditions of the village, the Lutheran church, a functional rather than impressive structure built in 1925, is tucked away on a pleasant tree-lined street. One other institution of importance cannot be missed. The barnlike building at the east end of town, the Turner Hall (home of a traditional German-American athletic club), provides the most obvious link between the contemporary community and its independent German heritage.

These two small Iowa communities, then, though both have successfully warded off the detrimental effects of change, have traveled different routes to the present. In a small, self-contained community such as Auburn, the glue that holds the community together is easy to identify in the church and the adjacent school. In Kane, whose origins and population are more individualistic, ethnic pride, coupled with a determination to succeed, were the driving forces behind its community and growth.

Community Building in Auburn

No sharper contrast can be imagined to the evangelical Protestant tradition of an independent laity choosing their pastor and their style of worship, than the hierarchical establishment of the American Catholic church in the nineteenth century. In rural, isolated parishes, the appointed priest had considerable power and much latitude in carrying out his duties. Though there was room in which a system for abuses, the remarkable record of the two priests who presided over the Auburn parish from 1880 to 1969 shows how successful it could be.

Father Francis Boeding was appointed to Auburn from the seminary in Canada when the community was just emerging from the frontier stage. Photographs show him to have been an austere figure with an aristocratic face and bearing, seemingly born to command. As a believer in German cultural superiority, he set standards and defined the physical boundaries and membership of the community. In particular he sought suitable recruits for settlement in the early days, was active in arranging marriages between parishioners, and most important of all, remained on the lookout for land that could be incorporated into the parish landholdings. Unlike some communities, Auburn was not transported lock, stock, and barrel from the old country; it was an amalgam of German-speaking Catholics who had made their way to northeastern Iowa from all over Germany and Switzerland. It was Father Francis's task to blend this heterogeneous group into a cohesive parish. His stature outside the community must also have been considerable, for although German was spoken up to the twenties, the village suffered none of the German baiting experienced in other Iowa communities during World War I. On occasion Father Francis went out of his way to assist other neighboring Catholic parishes. Once he personally lent another congregation money to make their mortgage payment. His major achievements, however, were the germ of educational excellence he instituted in the elementary school and the tradition he established of sending bright parish children to seek vocations in the church.[16]

At Father Francis's death in 1928, Father Bluh began an equally long and illustrious tenure. His first task, by order of the bishop, was to Americanize the parish. Slowly, under his guidance, English began to be spoken in the school and in the home, and sermons were no longer given in German. Bluh broadened the educational commitment begun by his predecessor and also continued to encourage families to seek land beyond the boundaries of the community. These policies were designed to give children the chance to farm if they wanted and to open opportunities in other occupations in case they did not. The certification of a high school in 1941 enabled those children who wanted nonfarm occupations to obtain a diploma without leaving Auburn. Gradually in these years the community grew less isolated.[17]

Community Building in Kane

Community building in Kane was more tenuous. The earliest settlers identified not with the church but with their common Holstein heritage. Then the railroad founded a corporate-sponsored community, some of whose residents had nothing in common with those on farms outside village boundaries. The core group of twenty or so families who had homesteaded the prairie by the early 1870s spent most of the first two decades raising families and working hard to improve their farms. Indeed, Kane lacked institutional development, apart from the mandated country schools. The first school in the village did not appear until 1887, when the railroad donated land for the purpose. Although a small Catholic church was built in 1870 to serve the Irish Catholic farmers, the foundation of the Lutheran church (Missouri synod) actually postdated the establishment of the major recreational institution in the village, the Turner Hall. The Turner club was initially a town organization, and the first board was recruited from the ranks of the town dwellers. The Lutheran church, on the other hand, seems to have been farmer sponsored. The signers of the church constitution in 1895 were mostly from prominent farm families. Services were initially conducted in the village school by a pastor from a neighboring community.[18]

The casualness with which the major institutional building blocks were founded in Kane emphasizes the world view of Holsteiners in the late nineteenth century, which to a certain extent continues to influence the community today. Originally, Holsteiners came to America determined to jettison the interference in their lives from the state and the state-sponsored church, which they had experienced at home. They resolved to use the opportunities available to them in America to better themselves, to own land, and above all to prosper materially.[19] While farming was a way of life for them, they were also progressive and very professional when it came to the agricultural development of their community. By the turn of the century, when the pioneer generation had married off its children to neighbors and begun to hand over its land and retire to town, the foundations had been laid for the prosperity that would blossom during the golden age of agriculture.

Kane farmers and village dwellers had no spiritual blueprint for their lives. Like good midwesterners, they were mainly preoccupied with material success. If they farmed, they felt obliged to set their children up in farming in the style they themselves had enjoyed. Because of the close proximity of a number of rival communities on the railroad and on the Northwestern main line to the south, there was considerable pressure for residents to support the village institutions and to keep up appearances. By the turn of the century, the gradual homogenizing of the community made these reasonably easy tasks.

By 1910, when the first class at the new high school graduated, the village of Kane was prosperous enough to support a hotel, a music teacher, and its own electric plant. From the turn of the century, the Turners, not the church, provided the community with its major cultural and social diversions. Unique in America, the Turners promoted serious interest in gymnastics for youth and also provided an active social program for all members. To this day, though most Turner clubs closed their doors long ago, the Kane branch flourishes as a social club. Its gymnastic activities have been replaced by bowling. Appropriately, the Turners still sponsor the ancient German ritual of the Birdshoot, held each year on the first Saturday in June. Here blindfolded children attempt to shoot a wooden bird from a high pole. The festivities are marked by a parade, sideshows, and dancing in the evening.[20]

Community leadership in Kane is diffuse. When the banks went under in the Depression, it was in the interest of farmers to ensure that a new institution was built. In a short time shares were sold, and the board, composed mainly of farmers, opened a new bank. There is a general consensus about the maintenance of community standards, which is fueled by pride in Kane. Although most residents probably had the same kinds of doubts about book learning as other rural folk, they realized that good schools were essential for the sake of appearances. Hence there was relatively early investment in a high school, supported by farmers and villagers alike. By 1926 half the graduates of the high school were male, and by 1930 many sons of the farm families were also graduating. A successful high school, especially in sports, was vital to the town's image and to the maintenance of a viable community. It is no accident that Kane voters over the years have a perfect record of support for school bond issues, which have systematically modernized and updated facilities.[21]

Community in a Changing World

Given its limited resources, Auburn made great strides around the time of World War II to improve the opportunities of young people. The establishment of a high school whose primary aim was to train children for college entrance was built upon the tradition of preparation for a vocation in the church. The other strategy of encouraging families to buy up any available land in the surrounding area was not, of course, officially church sponsored. Nevertheless, the local priest continued to press for these purchases, which allowed many young people to remain in farming at a time when opportunities were limited in many other areas. The continued secularization of the community and the movement away from prewar isolation were assisted by the involvement of many young men in the war itself. The most rapid change in life-style occurred after the war when former soldiers, who had seen how the rest of the world lived, would no longer tolerate the primitive living and working conditions their parents were used to. Flush toilets, central heating, modern kitchens, often installed by the men themselves, rapidly transformed living conditions on the farm. At the same time, the rapid mechanization of farm work itself, especially in the milking barn, made a great difference to the daily round.[22]

Over a twenty-year period down to 1960, 228 students graduated from the Auburn Catholic high school. During these years the bulk of the teaching load was absorbed by the priest himself and eight teaching sisters. In the early sixties increasing costs and lack of available staff forced the school to look for assistance from lay teachers, who, though they worked for next to nothing, were too costly a burden for the school to bear. In 1967 the high school graduated its last class, and two years later the elementary school was forced to surrender the seventh and eighth grades to the junior high school in the new consolidated district.[23]

There were many ironies in this experiment with an autonomous high school in Auburn. The first was that this little high school was able to provide first-rate college preparation. A glance at the high school year-books and genealogies shows that many of these children went on to successful careers in many areas. On the other hand, those students who were less academically gifted had little pressure to attend high school; even when it was mandated, they could easily drop out to farm. Unfortunately for the community, the timing for a decision on the reorganization of their high school could not have been worse. It came when their priest of forty years lacked the kind of energy needed to deal with a crisis situation. In any case it is doubtful whether the expenses of an accredited, consolidated Catholic high school could have been met, given rising costs. Fortunately for the community, the elementary school was saved through consolidation with another parish, and children continued to receive religious education from kindergarten through the sixth grade.

From the late 1960s, Auburn children began to attend the consolidated public school at Deer Valley from the seventh grade on. The district, which at its inception had roughly 90 percent Catholic enrollment, was an analgam of five small communities to which Auburn families had traditionally exported their young men and women to farm. Interviews with Auburn residents reveal some unhappiness with academic standards in the past few years. The high academic achievements of Auburn's brightest children when the town had its own high school were difficult to replicate at Deer Valley, partly because the communities involved had different expectations for their children. The Irish, Bohemians, and Germans had very different views of education. Moreover, it is important to remember that the standards set by the old parochial high school coincided with the highest level of achievement in American high schools. The consolidated school, however, operates during a time of stress for secondary education, when outside influences make teaching and learning difficult.[24]

Not surprisingly the emphasis at Deer Valley is on those activities that give the school maximum exposure and concretely demonstrate to taxpayers that their money is being well spent. In a totally rural area, with few competing attractions, sports takes on additional importance in community building and can even seem a panacea. Winning a state championship, as the consolidated school recently did, provides the community with a more concrete achievement than does excellence in academics.

As a progressive community, Kane believed in providing the best available educational facilities and opportunities for its children. Consolidation of the rural, one-room township schools occurred in 1946 and created none of the bitterness associated with such reorganization in many parts of the state. A more ambitious consolidation occurred in 1964, when five Milwaukee railroad communities and one other formed a single large district. The common heritage of the villages and the open country neighborhoods suggested that such a combination might be successful. The loss of a school can damage the future viability of any small community, but in the bargaining that went on before reorganization, Kane was able to secure the new district's elementary school. The junior high and the high school went to other bidders. After almost twenty years of consolidation, rumblings of discontent came from one of the villages that had been left out of the redistribution of school plant.[25] Recent school board elections saw bloc voting in the dissatisfied community in an effort to unseat the at-large board members who have expressed satisfaction with the status quo.

Boom and Bust

The last years of the seventies and the early eighties gave Auburn and Kane a stiff test of their ability to survive in a world where small communities were especially at risk. The generally cautious way in which expansion took place ensured that both communities avoided the worst pitfalls of the economic downturn. But though the crisis took longer to reach them than areas to the south and west, eventually it enveloped the local farm economy on which both villages were so dependent.

In Auburn, the physical limitations of the terrain, together with a lack of farmland available for purchase in the surrounding area, caused those who needed to expand to look beyond the boundaries of the community for suitable land. Among these livestock producers, expansion was not for speculation but to provide greater self-sufficiency in feed. Therefore, when the downturn began to affect the farm community in the summer of 1985, it was less damaging to farmers than it might have been. The most vulnerable farms were those with a part interest in an activity related to farming, such as a tiling business. Some young farmers who tried to be independent of parental control were also vulnerable. In a time of uncertainty when lenders demanded that borrowers show cash flow, a hog-dairy enterprise had its advantages, because it permitted a year-round production schedule that, even at a time of poor prices, kept a farm family afloat. Auburn farmers, therefore, were able to fall back on well-tried strategies of hard work and frugal living, which had served their families for several generations.

In the hamlet itself, the seventies were a high point for the half dozen or so businesses that catered to the surrounding farmers and others farther afield. With the downturn, they retained their high standards and kept the loyalty of their clientele. In a conservative community like Auburn, where the farm remained a full-time occupation for most families and the church continued to be the center of community life, there was less temptation to indulge in the golf–winter vacation syndrome than in other places. Few farmers played golf, and few could afford to spend their summer weekends at the Mississippi River, which became a popular recreational ground for many northeastern Iowa farm families in the seventies. Ironically, by the middle eighties, despite the downturn, country club memberships and interest in golf had grown considerably. Regardless of their rigorous summer schedule some farm couples found time to escape to the club once a week.[26]

In the seventies the roles of the elderly and women also changed. For the "young old," those who were healthy enough to take advantage of the freedom of retirement, the seventies and early eighties were a stimulating time. Landownership, which until recently had been their principal source of livelihood, provided them with the wherewithal to enjoy retirement. Not only did several couples indulge in winter vacations in the South, but they also took trips to Europe to visit the birthplace of their ancestors. Retirement meant a move from the farm to the village for most elderly farm couples. On the other hand, it also meant that some remained in their town homes alone, often for long periods, after the death of their spouses. Children made every effort to keep in close contact with them, and only as a last resort would elderly parents be placed in a nursing or retirement home. This kind of care was expensive, and most wanted to remain at home as long as possible.[27]

Partly because of the village's isolation and because the rigorous routine on most farms gave farm wives a position of some importance in the rearing of livestock, the movement of women to work outside the home did not attain great importance in Auburn. An occasional farm wife worked as a schoolteacher or for county government, and some women were partners in village businesses with their husbands. For the most part, however, there was little pressure from women to broaden their lives, or from their husbands to earn money when times became tight.[28]

Thus, the prognosis for Auburn in the immediate future was

reasonably bright. The community would survive the credit crisis because farmers there had never indulged in reckless expansion. Yet the long-term effects of the disruption would be considerable, for the downturn made many young people think seriously about leaving the area. Despite their progressive record, the businesses in the village probably will also have trouble surviving, if for no other reason than that a further generation of managers will not be recruited to take over. An even bigger question mark hung over the future of those farms that depended on cooperation between generations and made up for its lack of sophisticated technique by hard work. Such operations, which had served Auburn so well in the past, might have some difficulty in the uncertain climate of corn-belt agriculture a decade hence.

In Kane, where attitudes and beliefs were rather different from those in Auburn, the impact of boom and bust was not the same. The village was somewhat larger and not nearly so isolated. Because it was within commuting distance, it was a bedroom community for a certain portion of the population who worked in Cedar Rapids. There was also a cliquishness about the community, which ranged old, established families against newer arrivals, who in some cases were not church members. The farm population continued to be dominated by a core of old families, whose heirs farmed on a modest level, raising either cattle or hogs or both and producing seed corn.[29]

As in Auburn the full force of the farm crisis took some time to reach the community. The tradition of cattle feeding was continued by a handful of families; it was they, with their requirement for borrowed money, who were particularly vulnerable when the national crisis hit the community. One early failure came through overexpansion because of a desire to bring family members into the operation. Others occurred when young farmers got overextended on machinery payments. In 1985 a group of "outsiders" determined to influence the community into taking some positive steps to solve the problems they saw in production agriculture. As early as February 1985 forty farmers attended a meeting to discuss new ways to approach marketing.[30]

In September, with the downturn getting worse, a community committee brought in speakers from the National Farmers Organization to discuss their programs at the Turner Hall. An overflow crowd of over five hundred people attended this meeting to hear NFO speakers explain how a locally sponsored program would fit into their national effort to raise farm prices.[31] The NFO Committee worked hard to persuade farmers within a wide radius of Kane of the utility of NFO marketing strategy. In January 1986 they achieved one of their goals of opening a hog-buying station just outside the community. But most of the "insiders" remained skeptical. Many agreed with the principle of trying to raise prices through collective action, but they were worried that an organization with the controversial reputation of the NFO would destroy the fragile business arrangements with which they were comfortable.[32]

Loyalty to the community and especially to businesses within the community was considered vital to the successful functioning of Kane. As far as possible business was conducted within the community through the grocery store, the three grain elevators, the John Deere dealer, the gas station, and the funeral home. Competition, except in the case of the elevators, was frowned upon, as the proprietor of the town cafe found out whe he tried to challenge the only tavern in town by staying open later than was customary and selling beer. Thus, the entrance of the NFO caused some disquiet, especially because one of its major programs depended on grain sales, which, though using local elevators, would disrupt well-established marketing practices. Although there were innovative minds within the farm and business community who were prepared to employ unorthodox methods to achieve change, the NFO was a little too adventurous for most.

A similar situation developed with the opening of the golf course and country club in 1985. Unfortunately for the organizers—a farm family long resident in the area but with ties to the local metropolis as well—this ambitious project had been planned when the economy was still booming. By the time the money had been borrowed to convert an eighty-acre field into a nine-hold course and to move an old barn and renovate it to serve as the clubhouse, the economy had begun to turn sour.[33] By cruel irony, this first-class facility, which symbolized the dreams and ambitions of rural America in the boom period, opened at one of the darkest periods of the farm crisis. Here was a facility that in the best of times could only operate at full capacity seven months in the year, competing head-on with the venerable Turner Hall as the center of social life in the communiy. Although the club drew members from several other nearby villages and did not depend entirely on the patronage of Kane, it remaiend a rather incongruous symbol of the boom, stuck out on the prairie. How well it succeeds in difficult times will provide a very interesting measure of the depth of the farm crisis in east-central Iowa.[34]

Compared to hundreds of other small Iowa communities, therefore, Kane could build on a number of strengths handed down from the past and applicable to the difficult times ahead. Though this homogeneous community, often had problems absorbing and tolerating the influence of "outsiders," and most of the ambitious plans were put forward by "outsiders," there were some advantages in times of adversity to having close ties within a community. At an institutional level the village had great strengh. The local savings bank, for instance, which was controlled entirely by community shareholders, was run on conservative lines. Loans were made in amounts well below maximums stipulated by state regulations, and therefore the institution had few troubled loans. On the other hand, the restaurant had continual problems down the years, changing hands repeatedly. Farmers and business people traditionally used it as a place to exchange gossip over early morning breakfast. Most of the clientele spent long hours nursing twenty-five-cent cups of coffee, which hardly gave the management a satisfactory return. Although the community wanted its restaurant, it apparently was not prepared to give it the necessary economic support. In 1986 the ultimate in "small-town socialism" overtook the business: the village itself assumed management for a trial period after private ownership failed.

In a one-church community, resources could also be mustered within this institution. One of the characteristic developments of the boom period was the work of the Lutheran pastor, whose powerfully personality had made a pronounced impact on his congregation. For years the church was treated more as a social institution where members went before dropping by the Turner Hall before lunch on Sunday. The arrival of a charismatic pastor not only helped invigorate the spiritual life of the congregation but also introduced family counseling on a regular basis. For stress in the boom period of the early eighties could be as trying for farm families as were the pressures of the downturn later.

In Kane the tradition of working hard and playing hard, the close proximity to a metropolitan area, and the attempts to display the cut and dash which symbolized success in the rural world gave the changes among women in the seventies a greater impact than in a more isolated community like Auburn. Farm divorces, which were unheard of in Auburn, occurred in Kane, and the pastor spent a good deal of time trying to save marriages. In the early eighties, it was common for Kane women to aspire to work outside her home. Some enrolled in the local community college, and according to the pastor, there was rivalry among farm families for the spouse to secure an off-farm position and bring money to the farm from the outside.[35] As the farm economy worsened, such jobs were no longer a luxury but a necessity for the farm family. Unfortunately, even minimum-wage part-time jobs, such as nurse's aid at the health care center (the largest employer in the village) or cook's helper at the school, were scarce.[36]

In sum, Kane though by no means free of many of the common ills that afflict small communities in a downturn of the farm economy, had a promising legacy on which to build for the future.

Central Valley Community Building

Early promoters and residents of many Central Valley communities wanted to give their towns the appearance and the ambiance of midwestern country towns, and they were largely successful. Only the palm trees gave them away. Yet, despite appearances, the social structures of these towns never had much in common with their models in the Midwest. Hidden away from the main thoroughfares in Hanford was Chinatown, and south of the railroad tracks were shanties, labor camps, and brothels, where the underclass lived. The climate also contributed to the kind of society that evolved in the Central Valley. Mild winters attracted settlers from parts of the country with a frostier climate. At the same time, the summers, with temperatures often rising above a hundred degrees in July, were too hot for many to work outside.

For the first three decades of the century, Hanford was dominated by a resident elite whose wealth was derived from agriculture and landownership. These were sophisticated people, who lived more like the planter class in a colonial territory than like prosperous midwestern farmers. The family heads were gentlemen farmers who often hired ranch managers to look after their everyday affairs. Because fruit culture was an important component of the economy, and the marketing of fruit called for packing plants, alliances were formed between ranchers and agribusinessmen. Before the Depression, the Hanford elite lived a leisured existence. They built ranch homes north of town and often had houses in the city also, especially if they had high-school-age children. Hanford's ideal location, only a few hours from the San Francisco Bay area, the Sierra, and the central coast, allowed these families to take advantage of the social, cultural, and educational facilities of the rest of the state. The pre-Depression elite helped to give the community its polish. This legacy is still apparent both downtown, where several impressive buildings were completed in the twenties, and in the rural areas to the north of the city, where some old ranch homes have survived. Oil and gas, which began to be exploited in the Kettleman Hills on the west side in 1929, gave promise of economic diversification, and in the early years of the Depression, oil took some of the sting out of the downturn.[37]

Nevertheless, during the thirties the old elite faded away, and their places were taken by families that, though still connected to agriculture, came from more modest backgrounds. The thirties also saw the revival of the Democratic party in the county, sparked by younger men who would remain influential in local politics until the sixties. Their New Deal leanings pricked the conscience of the community and led to acceptance of immigrant farmers, who until then had remained outside the mainstream of town and county social life.[38]

As the seat of a California county, Hanford would feel the impact in the forties, fifties, and sixties of federal and state governmental intrusion. California was as successful as any state in attracting federal dollars in the form of military installations. Great efforts were also required from local officials to secure such facilities, and Kings County lobbied hard for the Naval Air Station at Lemoore, which opened in the late fifties.[39] Other federal activity included the irrigation projects of these years, and the concomitant rationalization of county water districts added a new layer of bureaucracy to the gradual proliferation of county government. It also augmented the white-collar work force of Hanford. Throughout these years, agriculture remained the largest employer, with government and then the military close behind. Thus the character of Hanford gradually changed. In appearance it remained a "stable mature yet modern city," far more attractive and sophisticated than midwestern county seats of comparable size, but employment became more and more diversified, and a large proportion of its residents had little or nothing to do with agriculture.[40]

The idea of luring a large federal military facility was broached by a county board dominated by representatives of the small farmers around Hanford. Lake bottom agribusinessmen were initially opposed, because they feared that a larger nonagricultural employer would force them to raise wages. The Navy Department's criteria for site selection included perfect flying weather throughout the year and large amounts of open space around the airfield, so that the site would not be subject to intrusions from urban sprawl, as had been the case at their facilities in the San Francisco Bay area. The county board was able to convince the navy that these needs could be met and that the county was capable of hosting an extensive military installation at a site on the Fresno line near Lemoore. Although the expectations for growth, particularly from real estate, did not materialize, the influx of thirty thousand military personnel and their dependents, many of whom lived off the base, had a major economic impact on the northern part of the county and helped diversify an agriculturally dependent economy. Similar efforts to attract industry to Hanford at this time were reasonably successful as well. A large tire-manufacturing plant, an oil refinery, and some agricultural processing plants provided new employment in the sixties and seventies.[41]

One of the major changes to occur in Hanford resulted from the expansion of local government in California in these years. County government moved from a downtown courthouse to a new "campus facility" to the west of town. The new courthouse was a monument to

Hanford's desire to keep the county and city progressive in comparison to other Central Valley communities. The move left a number of architecturally significant buildings, including the old courthouse and jail, vacant. Their eventual renovation by an entrepreneur, assisted by government grants, symbolized the distance Hanford had traveled from its agricultural past.

The idea that Hanford could sustain something approximating tourism was perhaps a risky proposition. On the other hand, for years the town had supported one of the best restaurants between Los Angeles and San Francisco, so there was a precedent for local support for up-scale consumer taste. Essentially what developed in downtown Hanford in the eighties was a tiny version of the kind of retail development common in tourist communities on the coast. Hanford saw the transformation of the courthouse into restaurants, bars, and boutiques and the conversion of several Victorian houses into a country inn, whose standards for its bed and breakfast business were as high as anything in other areas of rural California.[42]

Dilemmas of a Rural Elite

The influx of well-educated business and professional people from other areas of California and the rest of the country has changed the character of the community in the past twenty-five years. Leadership became more diffuse, and it was difficult to detect "leaders" either in the city or county. The relative complexity of the local economy and the social structure of the community rearranged the position of farm families in the configuration of community organization. To be sure, a number of younger farmers bridged the gap between the ranch and the city and achieved some business success in both spheres, but by and large, farm families were more isolated from the mainstream of community life than was the case fifty years before.

Partly, this isolation was attributable to the ethnic organization of farmers in the area surrounding Hanford. Established ethnic communities like the Dutch and the Portuguese had their own voluntary associations, churches, and even schools. The rural social structure was further modified by the migration of urban dwellers to the country. This gentrification process, which took place mostly in the sixties and seventies, tended to remove what had been farm laborers' housing from the rental market and set ranchers and landowners physically apart from their employees. Although milkers still lived close to their employers, most agricultural workers lived away from the ranch and commuted to work. Once labor camps ceased operating, temporary workers lodged in town during prolonged visits from Mexico.[43]

As good an indicator as any of the changing status of farmers was the reception of farm children in high school. Forty years before, farm children, regardless of ethnic background, had great adjustment problems in Hanford High School. The rural elementary schools gave them poor preparation, and there was considerable prejudice by "townies" against ranch children.[44] In the contemporary period, the general affluence of farmers and their status as a rural elite have made the farm children a privileged minority. The Hanford High School Future Farmers of America had a deserved reputation in the Central Valley for excellence, much of it due to the efforts of parents who donated both time and money to ensure that their children received maximum benefits from the program. It was not unusual for children to display their animals at livestock shows in metropolitan areas all over the West, and parents thought nothing of transporting children and animals great distances.[45]

Certainly the motivation for this parental interest was to ensure not only that proper socialization took place at home and at school but, more important, that the child's extracurricular activities were worthwhile. Farm parents, afraid that children would succumb to bad influences, walked a fine line between overindulging their children and being too strict. Farm families, whose earlier generations had worked so hard to establish themselves, found it difficult to instill a similar ethic in the younger generation. And the sensational Matt Slayton murder case, which merited a twenty-column story in the Fresno *Bee* many months before his trial began, seemed a fulfillment of these fears.

The details of the tragedy can be briefly recounted. Eighteen-year-old Slayton, the grandson of a self-made millionaire rancher who came from Arkansas in the thirties, contracted the murder of both his parents by three of his "Continuation School" acquaintances. This they achieved just before Christmas 1983, and all four were quickly arrested. Over the next eighteen months, Slayton and two of his accomplices—one of the conspirators had turned state's witness—were tried and convicted of the murders.

The case is interesting for what it says about the youth culture of small cities in the Central Valley, about the relationship of the nouveau riche to their children, and most important the moral of the case for the community and particularly for conservative farm families struggling to shield their children from undesirable elements in school and town.

Teenage life in the southern San Joaquin had not changed much from the fifties. Automobiles held a special place, and the most important status symbol was spokes on the wheels and raised white letters on the tires, which cost as much as twelve hundred dollars. A good night out would consist of "a few beers" and "a cruise" up and down the streets of Visalia or Hanford. The children of farmers especially favored fifteen-thousand-dollar Trans-Ams or eighteen-thousand-dollar four-wheel-drive pickups, and when they finished cruising, they liked to concentrate on "partying and dates." It was this car culture that allowed Matt Slayton to recruit his accomplices. They were far less well off than he, and the ten thousand dollars he promised to give each of them from his inheritance would buy them their own sports cars.[46] The close-knit teenage world of Hanford proved their undoing. Everyone knew everyone else's business, and a number of youths whom the police questioned after the murders claimed that Matt had boasted that he would kill his parents for his inheritance.[47]

According to their friends, the Slaytons wanted their son to be an all-American boy, but by the time he was fifteen their relationship had soured. The father was a fanatical worker, but Matt wanted nothing to do with the farm. To try and win their son over, the Slaytons used a carrot-and-stick approach, lavishing sports cars and stereos on him and then taking them away when he transgressed. Once, Matt's father threw all the nuts and bolts in his workshop on the floor and forced his son to sort them into their correct boxes. On another occasion Matt slashed the tires of his sports car and hit the engine compartment with a sledge hammer after his parents withdrew driving privileges.[48] The estrangement was so severe that Matt left home to live in Hanford while he went to high school. Transferred for poor performance to the Continuation School (a school for dropouts), he fell in with youths from less privileged backgrounds, and there he made plans for punishing his parents.[49]

Although Slayton's accomplices were all working-class Anglos, they had more in common with him than many of the youth living in Kings County, for the position of the landowning elite in the countryside was accentuated by the Hispanicization of the county from the sixties onwards. Roughly a quarter of the population was Hispanic by 1984, and a fifth of those with school-age children were on Aid to Families with Dependent Children.

The one-room country schools in Kings County had been consolidated between 1940 and 1955, but elementary schools, grades kindergarten through eighth, had remained independent institutions with their own boards. Meanwhile, private schools became more and more popular among Anglos, and soon it was possible to find an incongruous situation where a school with a majority Hispanic enrollment had a board dominated by farmers whose children attended private schools.[50]

The temptation to turn their backs on the outside world became increasingly strong for affluent rural Anglos. Fundamentalist churches were in the forefront of the movement to supply a cordon sanitaire between their children and the rest of the population. There were half a dozen church-sponsored Christian schools in the county, including two that took children from kindergarten through high school. Their popularity was a symptom of the changes in rural areas in the past half century. Whereas the rural elementary school was once the center of neighborhood activity, now schools reflect social class and ethnic origin, and there is little contact between different groups. Such trends were not unique to this rural area but represented the spread of a universal characteristic of contemporary urban life. In the rural area of northern Kings County, they signaled the decline of community based on the concept of place but a strengthening of community ties made through the ethnic group and church denomination.

Water as Community Builder

The importance of water in the Central Valley can hardly be exaggerated. A major confrontation over water rights that occurred between members of two communities analyzed here provided a kind of bridge between them and helps to illustrate the different organization of farming in the two areas.

Because it was not possible to pump groundwater from the lake bottom area, local farmers had for years searched far and wide for sources of water both above and below ground. All over northern Kings County water rights were acquired by the lake bottom ranchers from landowners lulled into believing that supplies of water were inexhaustible. The purchase of stock in mutual ditch companies was one perfectly legitimate method of gaining control of a reasonably secure supply. Much more controversial was the purchase of groundwater rights, which could only be utilized by pumping water to the surface and then transporting it some distance to a storage area or to the point of irrigation. In effect, such a procedure robbed Peter to pay Paul. Heavy pumping lowered the water table not only on the farm where the water rights were purchased but in the surrounding neighborhood as well.

In 1951 there was a direct confrontation between the lake bottom interests and the smaller growers in the area east of Hanford in Kings and Tulare counties. Most small operators lived away from the river and ditch delivery systems and depended entirely on groundwater for their needs. As a reaction to underground pumping, a group of farmers established an organization they called the Water Protective League. Their objectives were to halt the pumping through a class-action suit and then to promote the formation of a public water district to oversee water management in northern Kings County.[51]

The first imperative was an injunction to stop the pumping. The lake farmers had sunk two types of wells, shallow wells of four hundred feet or less and deep wells of a thousand feet or more. When pumping drained the underground water from the upper stratum of rock, they shifted to the deep source, which was too expensive for the average farmer to tap. In May 1951 the Payne Company began pumping water around the clock. Their pumps were so loud that they kept the whole neighborhood awake at night. According to one of the principal organizers of the Water Protective League, the water level dropped twelve feet in two weeks, and he was forced to spend a large amount of money irrigating his crops. The pumping of underground water, he said, was the equivalent of coming onto his land and driving his six cows away. Water was so vital to ranching "that land without water in Kings county was worth nothing." If the small farmers failed to control their surplus water supplies, he concluded, "we shall lose the land of our heritage."[52]

The lake farmers did not see their pumping in this light, however. They hired their own San Francisco attorney to protect their interests in court and at county supervisors' meetings. The level of mistrust between the parties was such that it took ten months of wrangling before the supervisors were able to agree to place the issue of whether a water district should be formed on the ballot. One of the reasons for this delay was that the lake farmers themselves had aggressively pursued a case of their own in the courts. Charles Wood had filed a lawsuit in Tulare district court charging the Water Protective League with the same underground pumping violations the small farmers had brought against the lake interests. A total of 545 farmers were named as defendants in this action, which was described by a Water Protective League attorney as a blatant diversionary tactic and an attempt to secure a more favorable position as plaintiffs rather than defendants in court.[53]

In order to obtain a flavor of the passions the dispute aroused and an understanding of the community dimension of the confrontation, it is worth quoting at length from an editorial that appeared in the local newspaper:

> What is the so-called "Lake Group"? Is it a tight-knit organization of people, living beside or near a lake, holding meetings, and plotting destruction? Or is it more accurate to describe those persons who live in what is known as Tulare Lake as a heterogenous collection of very large, large, and medium sized farmers and landlords. And what is the Hanford group? Is it a large segment of downtrodden farmers eking out a pittance on small acreages, bound together by misery and goaded into a state of panic calling for desperate action? Would it not be more accurate to describe the latter as a fairly large number of small and medium-sized farmers who own acreages throughout the

northerly portion of Kings County, and who also make up a typical sample of the county population?

It was important to remember that all parties were first and formost, citizens of Kings County. Their prosperity is all linked by the same needs—the profitable cultivation of the land, enough water, reasonable taxation, the right incentive to produce more, and above all a spirit of cooperation. There is room in the county for big farmers and small farmers. We must remember that in some parts of the world, 320 acre farms labeled small would be considered very large indeed, and land running into thousands of acres be considered empires. It is all relative, for large operators not only afford employment to many people, and add purchasing power to the community, but also pay a good percentage of the taxes to carry on the government. The small agriculturalists are the backbone of the community. It is upon their shoulders that the prosperity of the area rests. No business in any city will continue without the support and patronage of the smaller landowner who comes to town with his shopping.[54]

Despite this plea to come to terms and allow conservation to begin, the name calling continued for at least another year. The lake farmers charged the county supervisors with conflict of interest, claiming that they were sympathetic to the Hanford interest. But eventually in late 1952 the county board did vote unanimously to initiate an election to form a water conservation district in the northeastern part of the county. Almost simultaneously the lake bottom farmers drew up plans for water districts south of and adjacent to the proposed northeastern district. At the election, finally held in February 1953, an overwhelming 96 percent of voters favored the formation of a new legal entity in the northern part of the county to negotiate water rights and to nominate a board to oversee these activities.[55]

Even after thirty years, resentment still smolders about this episode. The confrontation drew battle lines that have persisted in dividing the county. Hanford area farmers resented what they considered the arrogant behavior of the lake farmers, both in pumping the groundwater and in trying to undermine the formation of a water district. The small farmers were too angry to appreciate the desperate need of the lake farmers, who operated in so risky an environment. If they ignored cooperation and community mindedness, it was because their economic well-being was threatened.

Nevertheless, despite these animosities, some real progress was made in water management. The establishment of a water district rationalized and stabilized the delivery of water. By regulating the level of the water table in dry years, the district placed the groundwater situation on a stable footing. In addition, like all other districts, it acted as a negotiating agent to purchase surface water from other districts. Thus when the Pine Flat Dam opened, it played a crucial role in negotiating the purchase of surplus water for members. Another service was the monitoring of sales of ditch stock by private individuals in the district. Whenever possible, the board would pressure sellers to make these valuable commodities available to the organization.[56] Thus the formation of a water district with competent management and concerned members furthered a sense of community among the farm families of the area. Ironically, in future years it became common for the Kings County Water District to purchase water from its next door neighbor, the Tulare Lake District, which was formed at the same time to serve the needs of the lake bottom farmers, who had tried to deny the rights of their neighbors to the north.

Community Building in a Company Town

The water-pumping incident provides a backdrop and an introduction to the fourth community analyzed here. Like the lake beside which it was built, Corcoran is unique in California. At certain levels it was successful, while at others it failed in its efforts in community building. Since the late thirties, Corcoran had a reputation of being a company town where the line of demarcation between rancher and employee was rigid, and relations were strained. Though these kind of tensions die hard, the economic realities of agribusiness are such that in recent years there has been softening of attitudes. Mostly this relaxation came about because the largest employer, B.J. Payne, adopted a more sophisticated policy of management and employee relations, and others followed suit.[57]

Because of the dangers of flooding, Corcoran is a more compact and integrated community than might have been the case had growers been able to live on ranches in the lake bottom. Furthermore, the wealthiest families learned to play down their affluence by understating it in the community. This was not always the case, for in 1947, under the headline "Valley Hay Barons Live in Luxury and Entertain in Style," a San Francisco gossip columnist spoofed the Corcoran ranchers:

> Flat cotton and grain fields. Treeless, almost brushless cattle ranges. No really gaudy hotels, theatres, or resorts in the little towns of the region. Which is probably why the local bon ton...live regally, colorfully, and expensively. The women wear the most advanced styles, and the men—for whom nothing can be said in the category of advanced couture—drink the most expensive booze and drive the best cars. They can afford to. They make lots of money. They party a lot. The parties aren't at all Noel Cowardy. No polite small talk; just supercharged technicolor gossip, and belly laughs. No cocktails; just Bourbon and Scotch. No languid premidnight farewells; the parties wind up very late

and are usually topped off with a nocturnal swim. Then the survivors climb into their twelve-cylinder jobs, or private planes and head for home.[58]

Although some of the old ranch homes are still occupied, nowadays the elite live fairly modestly in neighborhoods that are not surrounded by high walls with electronically controlled gates and security guards. As Corcoran is a small town, poor people live in fairly close proximity to richer ones, and there is no internal segregation, mainly because a sense of social responsibility by the town's largest employer had decreed that all workers reside in Corcoran. Since it is quite possible to live elsewhere and commute to work, this policy had a major bearing on keeping the Corcoran population in balance. It ensured that the community did not degenerate, as others in the Central Valley did, into a slum housing no one but impoverished laborers.

Corcoran, like many other communities in the Central Valley, owed its existence to the railroad. The town site was where a spur from the main Santa Fe tracks went east. It was laid out around 1905 by pioneers who saw the potential of the lake bed for bonanza farming. They in turn attracted other ranchers and investors, many of them from southern California. After the flood of 1906, the first attempts were made to build levees on the lake bottom. Initially, wheat was the main crop planted, but just before World War I, some ranchers experimented with sugar beets as a major crop. A million-dollar processing factory was constructed, but other uses had to be found for it when it was discovered that sugar beets did not grow as well as was hoped. By 1914, at its incorporation, the town had roughly a thousand inhabitants. Early pictures showed a dusty main street lined by a collection of stores, bars, and a hotel, with the Presbyterian church off in the background.[59]

The town's most illustrious early citizen and institution builder was the rancher J.W. Guiberson. He not only helped found the first church, the first bank, and the high school but also assisted in the experiments with cotton and in persuading B.J. Payne to build a cotton gin in the town in 1924. By that time many of the families still important in the community had begun farming and were starting to expand by buying out the small farmers.[60]

The advent of the Payne Company proved a turning point in the town's history. The company's involvement in financing and processing cotton between the wars and its later purchase of vast acreage on the lake bottom ensured its involvement in every aspect of town life. The Payne's were extraordinarily secretive about their business and family affairs. For example, during the bicentennial celebrations in 1976, every important family in Corcoran allowed the local newspaper to publish an article on its past achievements. The Payne contribution was the smallest, and it was confined to a few family snapshots.[61] That they could not sanction a more revealing article in the community newspaper, in which they held a controlling interest, suggested a particular paranoia about publicity.

From the earliest days the family had important connections in the state and, later, nationally and internationally. From their Los Angeles County headquarters they were able to see the big picture as it applied to agriculture in greater clarity than would have been possible had they lived in Corcoran. Water was continually an issue, but labor, and the intervention of the state and federal government in social and agricultural affairs, had recurring importance.[62] Over the years the Payne Company became a past master in using government programs to advantage—some called it farming the government. The official records of Kings County are full of mortgage transfers from Payne to the Land Bank in 1929 and 1930.[63] In those years a majority of Payne-financed cotton customers were in trouble. So for the first time the company took advantage of a government program, a Land Bank restructuring policy. It was, however, labor that had the greatest impact on the social structure of the town, and Payne would be a strong influence in this area.

Ranchers in the Corcoran area suffered an early trauma in the form of a labor conflagration in their own backyard. The cotton strike of 1933 remains the largest agricultural strike in American history, and it occurred during a time when the ranchers were undergoing severe economic difficulties. Not surprisingly, they vowed never to be embarrassed by labor again. That year they founded the Associated Farmers to curb labor militancy in California, and the Payne Company, both in the county and statewide, assumed a leadership role in the new agribusiness organization. Although the evidence is scanty, it would seem that throughout the thirties Paynes remained in the background, while others were more public in their antilabor stance. The Paynes evidently helped finance the Kings County chapter of the Associated Farmers, but others held the offices.[64] By World War II, the Payne Company had perfected its labor policies and image, which, for the time and place, could be regarded as progressive.

The best evidence with which to explore the labor question comes from hearings conducted at Corcoran by a state committee appointed to investigate agricultural resources in the Central Valley in the summer of 1950. At a time when mechanical pickers were beginning to take over from field labor, committee members were interested in the impact of structural change on a community whose labor force already suffered long periods of idleness. The Payne Company did not furnish a witness to testify at these hearings but instead sent in a long, detailed memorandum outlining company policy on the stabilization of employment for its workers. According to Payne, the laborer found it hardest to get work in February and March. The company made it a policy to rotate such work as there was in order to reduce hardships for families. Family heads who could find work on other ranches were free to do so and could remain in the company camp free of charge, and with utilities supplied. Apparently the company also provided food to those who lacked resources. Moreover, conditions in Payne camps were better than in many others. Large families were allowed two cabins; running water for cooking was available, as were hot showers. Cleanup men were regularly employed to keep the camps free of disease; both county and state health officers had access to the camps. The company also made every effort to encourage school attendance. Some of the most flagrant labor abuses in California labor camps involved the method of payment. Often laborers were docked for groceries and other items at source. Payne workers, however, were paid in the most efficient way possible and were not forced to stand around for hours waiting for their wages. No liquor was sold in the groceries, and the practice of charging 10 percent for converting cotton tickets to cash was discouraged by company stores.[65]

The Payne Company made sure some of their laborers sent testimonials to the committee. Cotarino Cardinas wrote a fairly typical letter. "The living quarters here on Payne camp #1 are very suitable," he said. "We get good service of everything. Plenty of water, lights and natural gas. Plenty of hot and cold water in showers. Camp is kept clean as possible. Cabins well repaired. I have lived here for quite some time, and like it very well."[66]

This picture of the Payne camps was undoubtedly somewhat prettified, but it illustrates how much care the company took to deflect criticism and let others bear the brunt. As the testimony of a labor contractor showed, relations between ranchers and workers were generally poor. "First of all," the contractor pointed out,

> the farmer has failed to maintain a proper and decent place in which his workers live. Ever since labor camps were set up... proper supervision wasn't enforced and as a result you had gambling, drinking, and other forms of vice before the very eyes of the children who had the misfortune to be raised under these conditions. I believe some of the conditions can be checked and improved if the camp owner would take a little interest in this respect and not only have his camp as a place to get someone with a hoe and a sack. It is true that most migrants have little or no respect for private property, especially their children, and continual damage to cabins, buildings and other private belongings is done.... As it is now, absolutely nothing is provided for recreational

activities and naturally the kids spend their time throwing rocks at buildings instead of on the ball field.[67]

The prejudices of a decade before still existed in the early fifties. The idea that migrant labor was shiftless and irresponsible and abused the welfare system was universal among growers. One was especially colorful in his description of camp behavior. In his camp, the workers "all get drunk and spend their money and whip their wives, and just raise cain in general." He noted that welfare "chiseling" was common and told of one man who had gotten seventeen dollars a week from the welfare department, while his wife and six children were back in Missouri. He believed that alcohol was behind the migrant labor problem. "Just as soon as dark comes," he said, "there is what we call liquors." They came from all over the valley, to take the money away from the migrant laborers: "If a fellow won't get drunk quick enough they see he has enough to get drunk. They roll him. They take it away from him. Crooked dice and crooked cards and prostitutes. Any way to take the money from him. A fellow gets drunk, then he is always thinks he can shoot craps a little better than anybody else. If you take alcohol away from them, you won't have any trouble with the relief problem."[68]

This evidence brings out some of the dilemmas Corcoran faced in trying to build a community with some credibility. On the one hand, despite their crude notions about their employees, the growers did face huge frustrations in the logistical setup of their camps. It was not very surprising that whenever possible they tried to hire Mexican immigrants, who had no dependents and were reliable hard workers. Then the introduction of the mechanical cotton picker gave growers the technology to replace most hand labor, the elimination of the labor camps also permanently cut the customer base of businesses in Corcoran. The photographs of downtown Corcoran on Saturday pay night, with workers crowding the sidewalks and stores, attest to the kind of business done during the heyday of the labor camp era. For the town, then, the demise of the labor camp was an economic disaster, but the growers were relieved to be out of the goldfish bowl, no longer subject to the scrutiny which the town's camps and school system received from state and county government. The normalization of the town's social structure also brought many residents of the camps into the town itself, and for years, Corcoran was noted for its shanties and the "blighted appearance" of certain sections.[69]

The elimination of hand labor in cotton picking made the acute seasonal unemployment worse than ever. In this respect the Goldschmidt thesis, which posited that large farming enterprises breed poverty in the community, corresponded exactly to the Corcoran situation. As the years passed and farming became more and more dependent on machines, the city began searching for alternative employment opportunities outside agriculture to relieve the chronic joblessness characteristic of the town.

During the fifties, sixties, and seventies, Corcoran was forced towards more responsibility in regard to its less privileged citizens. The pace was largely dictated by the major agricultural employers, Payne and Wood, who zealously guarded against intrusions by local, state, and federal authorities. The county was already courting sources of nonagricultural employment, and the large lake bottom ranchers resented these efforts. In the late fifties they fought county attempts to develop a master plan every step of the way. Corcoran withdrew from participation, fearing outside interference.[70]

The notion that the town was the particular bailiwick of the large ranchers continued into the seventies. As late as 1977, a medical team from the National Institute of Occupational Safety and Health came to test workers exposed to cotton defoliant for possible neurological problems. Because the large growers refused to cooperate with the team, they left town without completing their research. The Wood airport story, in which the Wood family constructed a seven-thousand-foot runway too close to a state highway and then defied the Department of Transportation for several years over its legality, was another instance of the arrogance of the Corcoran ruling elite.[71] So involved were they with their own operational problems that they even behaved high-handedly with each other. In one famous instance in 1970, a senior Payne employee tried to breach a Wood levee with a bulldozer and had to be restrained by armed guards and the sheriff.[72]

The idea of providing public housing or of seeking grants from federal agencies to fund the kind of public-sponsored programs routinely utilized in other communities of similar size is a comparatively recent phenomenon. Apparently, fear of public housing was a legacy of the labor camp days; the townspeople firmly believed that poor people could not be trusted with "decent" accommodations. Once inside, they would systematically destroy them. For some years the town relied on private philanthropy to supply some of its largest capital improvements—the hospital, the high school football stadium, and the senior citizens center, for example. Much of the money for these facilities came from the Payne family. One innovative and laudable Payne program was a college scholarship scheme in which two Corcoran students each year were awarded full tuition grants to the colleges of their choice. By and large the recipients took full advantage of this opportunity, and some even came back to work for the town's major employers.[73]

Growers, seeking a more pliant work force, began employing

Mexican labor, hastening the Hispanicization of the area. Corcoran's poor reputation over the years did not encourage stability, however, and only those that had to remain in the town for any length of time. As was the case in the northern part of the county, the public schools lost a percentage of enrollment through "Anglo-flight." Many families left the community to live in more congenial surroundings, and many of those who were required to live in the town, preferred to send their children to private schools elsewhere. It was a sad change. During the fifties, even with all the human problems, the children of cotton baron and black cotton picker had played together on the same sports teams.[74] The sense of community derived from common participation in public institutions declined. Then, change was forced on Corcoran by nature itself. The flood of 1983-1984 devastated the community. The already high unemployment rate rose to 32 percent, causing a loss of $32 million in city revenues and a 40 percent drop in sales tax receipts. Several years previously the town had begun to attract light manufacturing concerns to its industrial park. An aggressive redevelopment agency managed to attract five new plants to Corcoran, with around three hundred new jobs. In 1983 the city, always so reluctant to seek federal funds, won grants worth over a million dollars for housing construction and rehabilitation and other capital projects.[75]

But one of the more remarkable events in the history of Corcoran was the wooing of the California Department of Corrections for a maximum-security prison in 1984-1985. Knocked to its knees by the flood, Corcoran conceived the idea of bidding for a prison to provide instant growth for the town. The record of the state of California for establishing prisons in modern times was abysmal. No new facility had been built for twenty-six years. Whenever the authorities selected a site and approached a community, they were thwarted by irate citizens who would not condone a maximum-security institution in their midst. Thus, Corcoran's enthusiasm was seen as most unusual in Sacramento. The state was able to build not one but two prisons in Kings County, at about the same time, prompting one wag to announce that the county would soon become the prison capital of the world.[76] Approval of the Corcoran proposal was particularly rapid; barely a year was needed to clear through the red tape in Sacramento. Certainly this quick passage was partly a tribute to the influence the power elite had at the highest levels of state government. At the same time, the eagerness with which the community was soliciting a prison facility allowed Corcoran to vault over ten proposals already in the planning stage from communities that were reluctant to play host to hundreds of hardened criminals.[72]

Of course, one wonders how deep the enthusiasm ran among the population of Corcoran. As with almost everything else that has occurred in the town in the past fifty years, the population at large had little to say when the decision was made. It would be ridiculous to assert that this strategy, and many like it, were dictated from the twenty-seventh floor of a high rise in Los Angeles County or in the offices of the large growers in Corcoran, but certainly in a company town the elected officials usually do the bidding of the important employers. Therefore, the prison had only a handful of opponents.

Conclusion

Clearly, there is a strong relationship between the farm population and the community that serves it, and equally clearly, the kind of farming practiced has a profound impact on the community. Case studies in Iowa show that, despite rapid structural change, small communities, which could be defined as communities of both place and group solidarity, have successfully adapted. Farmers and villagers maintain a vested interest in the key institutions in the community, and the practice of patronizing local businesses, churches, and schools has enabled some communities to survive even while others nearby failed. In the Central Valley, where a half century of distrust marred relations between large and small farmers, a small city with a tradition of family farming in the surrounding area weathered structural change more favorably than did a corporate-dominated town surrounded by industrialized agriculture.

Though the two California communities showed a tendency in recent years to drift toward segregation, group solidarity has also become more important. The more affluent families have laid less stress on social interaction in the greater community but have developed greater solidarity within their ethnic and church communities. Moreover, the leaders in Kings County have from the fifties onwards consistently shown a willingness to try to adapt the local economy to the realities of structural change. They have sought industry and public institutions with which to diversify an agriculturally dependent economy. There was far less of this sort of planning, in Iowa, and therefore, when the farm crisis struck, the small towns, so dependent on farming and agribusiness, suffered greatly.

8 Corn-Belt Crisis

Because of the complexities of the farm crisis in Iowa and the many actors involved, it is useful to provide a schematic chronology of events, while introducing the protagonists. The crisis went through five stages. As table 14 shows, the process began in 1981 during an inflationary spiral that gave little hint of what was to come. The next stage, from 1982 to 1983, was a phase of collective denial, when most of the farm community was apathetic to what was beginning to occur and those who did appreciate the trends were ignored. During 1984 and the first two months of 1985, lenders and borrowers began to confront each other, and farm advocates began to build a grass-roots movement to halt foreclosures and bankruptcy. By the spring of 1985 the state as a whole began to mobilize to deal with the symptoms of economic stress, seeking a resolution to the crisis, which could begin to be seen in 1986. Obviously, events cannot be discretely compartmentalized, but this schematization helps make sense of how the crisis developed.

In introducing the principal actors, it is important to remember that presiding over the whole drama, a thousand miles away in Washington, was the federal government and the Reagan administration. Ironically, while advocating free-market agriculture, Reagan presided over the most expensive farm program in history. In addition, as agriculture faltered, his administration was extraordinarily successful at deflecting hostility away from the president himself and towards such targets as the secretary of agriculture and the federal credit system.[1]

In the first three years of the crisis, Iowa politicians at both the state level and in the congressional delegation played only minor roles. Iowa in these years experienced two divisive elections, one for governor in 1982 and the other for senator in 1984. Terry Branstad, a conservative Republican, won the governor's race, but another conservative Republican, Roger Jepsen, lost the Senate race to a liberal Democrat, Tom Harkin. At the same time Iowa backed Reagan for reelection in 1984.

Table 14. Schematic Chronology of the Iowa Farm Crisis

Stage	Time Period
Inflationary spiral	1981
Denial, beginnings of deflation	1982-83
Confrontation, build-up of advocacy	1984-Feb. 1985
Beginnings of mobilization, mobilization	1985
Beginnings of resolution	January 1986

After November 1984 the state's farm economy fell apart, and a kind of power vacuum developed in which the state seemed to be leaderless, partly because both the governor and the senior senator, Charles Grassley, also a conservative Republican, took time to distance themselves from the president. By the spring of 1985 both had made the necessary maneuvers and had begun to speak out against Washington farm policy. In 1981-1984 the most visible actors were those farm families who, for one reason or another, got into trouble earlier than others; the farm advocates, who gave them support; lenders from banks, the Farm Credit System, and the Farmer's Home Administration; attorneys knowledgeable in bankruptcy law; and the state's most important newspaper, the Des Moines *Register*.

As agricultural economists pointed out, considerable differences in economic condition among farm families made the crisis harder to deal with. A frequently quoted table from data collected in 1984 indicates that older farmers showed the least economic stress; they frequently had debt-to-asset ratios below 10 percent. Younger farmers on larger farms were more likely to hold debt-to-asset ratios of more than 40 percent (table 15).

By the fall of 1984, the deterioration of the economy forced the Cooperative Extension Service to plan a mobilization of resources to help farmers. The program, known as Assist, offered farmers a computer-based financial analysis package and, beginning in February 1985, a toll-free telephone hotline referral service. In February 1985 the Extension Service and Iowa State University also hosted a rally at the Ames campus, which, while giving the issue of farm economic stress maximum national exposure, also eased some of the tensions generated in the state over the failure of the agricultural establishment to recognize the symptoms of crisis earlier.[2]

The Ames rally was a turning point. From then on, farm advocacy became respectable, and resources could be mobilized to deal with economic and emotional stress. A number of new actors moved onto the stage, including the governor, the legislature, the congressional

Table 15. Financial Condition of Iowa Farmers, January 1984

	Debt-to-Asset Ratio					
	0-10	11-40	41-70	71-100	Over 100	All
Operators	38%	37%	19%	4%	1%	
Assets	31%	42%	24%	3%	1%	
Debt	4%	39%	47%	8%	2%	
Average age	59	53	47	45	47	54
Average assets per farm	$503,000	$694,000	$745,000	$470,000	$217,000	$615,000
Average debt per farm	$11,000	$160,000	$383,000	$375,000	$262,000	$156,000
Average equity per farm	$492,000	$534,000	$362,000	$95,000	−$45,000	$459,000
Average acres owned	233	298	271	172	131	261
Average acres rented	121	189	306	382	198	193

Source: Iowa Farm Finance Survey, 1984, Iowa Department of Agriculture, Iowa State University and Crop and Livestock Reporting Service.

delegation, state welfare and mental health agencies, the state bar association, clergy, social workers, community colleges, and most important of all, grass-roots advocacy groups. By the spring of 1985, economic conditions were so bad that even those farm families in the most prosperous areas and with few debts were forced to recognize the potential dangers.

The Background

At the best of times the realities of modern agriculture have made it extraordinarily difficult for the young to start farming.[3] Expansion to bring a young person into the firm increased debt loads. Thus, both young farmers and intergenerational operations that had expanded to accommodate family members became susceptible to failure. Weakened firms were early victims. Corn growers, for example, succumbed when drought struck southern Iowa, and beef feeders in western Iowa bore the early brunt of low commodity prices in the world and national markets.

With agricultural well-being so tied to land values, the massive decline in equity between 1981 and 1985 had a major impact on every farm. Land lost 55 percent of its value in four years in Iowa. In one year alone between 1984 and 1985, the average worth of an Iowa farm fell 25 percent, or $114,000.[4]

In the boom environment the question of whether land would pay for itself tended to be ignored, but once deflation set in, land purchased with the expectation that it would increase in value now had to be supportable through the income it could generate.[5] In a deflationary cycle it was impossible to generate enough income to pay for land bought during a period of high inflation. An acre of land bought for two thousand dollars at an interest rate of 11.75 percent required $227.50 in interest payments alone every year. If the land yielded a hundred bushels of corn that could be sold for $2.50 per bushel, the farmer would stand to lose twenty-five dollars per acre even before other costs were calculated.[6]

The combination of overvalued land and low prices spelled doom for those who had purchased during the boom and had to make payments with depressed commodity prices. Land values followed the path of commodity prices, and there were no buyers in the market to rescue plunging values. The aggressive farmer purchasers of the seventies were overextended and trying to save themselves, while investors were scared off from the depressed market. The obsessive substitution of machinery for labor also had a significant impact on the economic health of a farm family when the downturn came. Some farms had no particular problems with landed debt obligations but were forced into reorganization by their short-term operating debt.[7]

How, then, could a farm family absorb these shocks? During the Great Depression, as we have seen, families were able to weather hard times by lowering their living standards. When labor was a major input in agriculture, the willingness of farmers to tolerate low returns for their work measured their ability to survive a downturn. In the Depression era, labor amounted to almost 50 percent of all inputs on a farm, and even as late as 1960 labor was a more important input than machinery. By the 1980s, however, it was no longer possible to absorb deficits through lower labor costs. On most farms labor amounted to only 12 percent of inputs; machinery, feed, chemicals, and fertilizer now made up 49 percent.[8]

As in the Depression, the era of extreme austerity on the farm had an impact in many other areas as well. Tax bases eroded, conservation was neglected, and support for community institutions dissipated. Nowhere was the impact felt more than in the service communities of the Midwest, which depended on farming for their livelihood. The welfare of agribusiness suppliers, the local retail trade, and social institutions such as schools and churches, was directly linked to the prosperity and numbers of farmers. They, too, suffered in the downturn.[9]

Agribusiness firms, such as elevators and seed suppliers, which worked largely with unsecured credit, found more and more bad debts on their books. Write-down of these accounts was painful, and the Uniform Commercial Code made bankruptcies particularly disadvantageous for such firms. If a farmer defaulted on a loan from a lender with a prior security interest, the purchasing agribusiness firm was liable for the value of the products purchased. Finally, the greater the financial stress on other local institutions, such as banks, the greater the difficulty the small local agribusiness firm had in obtaining financing.[10]

The diversification of the Central Valley rural economy gave California a buffer against these economic shocks, but in Iowa the economy was completely dominated by agriculture. During the farm boom, retailing in Iowa was robust. Indeed, in the seventies some claimed that the state's economy was recession proof. Such judgments seem especially poignant in light of the failure, between 1980 and 1984, of roughly five thousand retail outlets, many of them in towns of under eight thousand people. Farm-related retail businesses, such as implement dealerships and hardware stores, were especially vulnerable. In the Central Valley the patterns for agribusiness firms were no different, with lower sales universal and bankruptcies commonplace. However, because of the larger percentage of people employed outside of agriculture, retailing in the southern San Joaquin cities and towns remained less affected.[11] In Iowa patterns developed that boosted larger communities at the expense of smaller. Customers tended to migrate to shop at better facilities in larger cities. In so doing, they further weakened the small farm-service communities.[12]

Some suggested that if these trends continued, farm survival would become dependent on off-farm employment by one or more family members. In Iowa, where off-farm jobs were scarce in the average small farming community and the downturn had eliminated many of the employment opportunities in agricultural machinery manufacturing and meat packing, such a strategy seemed dubious. Some intriguing data from Iowa showed that, despite the losses in retail businesses, permits for retail sales increased by five hundred to seven hundred every year between 1981 and 1985. While it is difficult to prove, it would appear that many of these permits were for part-time businesses conducted in the home by those who had suffered employment loss. Presumably a certain percentage of the new enterprises were being run by farmers, many of whom were still farming part-time. This impression is strengthened by data collected from Iowa farmers who left farming for financial reasons. One study of 482 ex-farmers conducted in 1984, found that only 13 percent had left the state, and 15 percent were unemployed. The majority had been able to find jobs close to home and, most important from the viewpoint of the social structure of their communities, had remained in their old homes with their families.[13] Interviews in rural northwest Iowa in the final months of 1986 confirmed that, whenever possible, ex-farmers remained in their old localities and tried to make a living as best they could.[14]

Without massive government intervention, however, the long-term prognosis for most small corn-belt communities was not promising. The specialized economic base and poor employment opportunities in rural areas suggest that a sustained farm depression would force many ex-farmers to leave their communities. Such an exodus would bring further retail decline, greater difficulties for agribusiness firms and banks, school enrollment problems, and the escalating disruption of community organizations and institutions.

The Other Voice of Agriculture

The "wonderful seventies" hardly prepared the farmers of the state, their lenders, agribusiness, state government, or the Extension Service for the wrenching experience of drastic deflation after 1981. Since optimism is the sine qua non for accomplishing anything in farming, it is not surprising that the farm economy was going through a sea change and that circumstances had altered the industry fundamentally. One organization that subscribed to what might be termed the neopopulist agenda for change in American agriculture—curbing corporate ownership, protecting the family farm, and promoting sustainable agriculture, localism in production and consumption, and production controls to raise farm prices—was the Iowa Farm Unity Coalition. The coalition lacked the ideological blinders that prevented the mainstream farm organizations from seeing the potential seriousness of the economic situation. Thus, it was in a position to furnish advocacy in political, legal, emotional, and financial matters to those farm families who requested assistance. And for over two years before the mainstream responded, it was one of the few organizations in the state to which farm families could turn for help.

The coalition was formed in January 1982 in a bank basement in Atlantic, Iowa, an ironic venue for the founding of an organization that would give banks so much trouble in the next few years. The new organization brought together farm groups, such as the National Farmers' Organization, the Iowa Farmers' Union, the American Agriculture Movement, and the United States Farmers' Association, with church groups, such as Catholic Rural Life and the Iowa Inter-Church Agency for Peace and Justice, and community organizations, such as Iowa Citizens for Community Improvement, the United Auto Workers, Iowa Citizen Action Network, and Rural America. The latter, which in 1985 would cut its ties with its parent organization in Washington and rename itself Prairiefire, provided the catalyst for much of the advocacy work in 1982-1985.[15]

The leadership of Prairiefire could trace its lineage to the so-called professional reformers, who became active in the social movements of the sixties and early seventies.[16] While serving a farm clientele, they were also concerned with educating the public about farm issues. Prairiefire organized and manned a hotline in Des Moines and remained the only easily accessible source of advocacy in the state throughout the farm crisis. In an area as large as Iowa, where some communities were over two hundred miles and almost four hours driving time from the capital, it was inevitable that Prairiefire was thin on the ground. Its staff was most skilled in the areas of political action and financial and legal advocacy. Therefore, it concentrated on these activities in counties close to Des Moines, and in the state legislature. Other affiliates, Catholic Rural Life and Citizens for Community Improvement, ran their own programs in other areas of the state and helped plug the gaps.[17]

As the "other voice of agriculture"—the Farm Bureau being the chief advocate of mainstream and conservative farmer opinion—the Iowa Farm Unity Coalition was formed, in the words of one of its founders, because "we were obviously entering a period of crisis, and most of the solutions proposed were about the same. Our goal was to provide a forum for analysis and strategy."[18] A major disagreement with the Farm Bureau centered upon the role of government in agriculture. The coalition wanted the government to provide a supply management program "in which production needs would be estimated and the total divided equally among farmers." Such a program would permit some land to be taken out of production for a season, thus conserving its long-term potential for growing crops. The coalition also advocated increases in government support loan rates to boost farm income and, until new rates took effect, a moratorium on foreclosures and bankruptcies. In short, the coalition considered current farm programs wasteful and ineffective and wanted the government to make wholesale changes to improve farm prices.[19]

These ideas, many of which were also found in the Harkin-Hightower proposals for the 1985 farm bill, were strongly resisted by mainstream farm groups from 1983 to 1985. In particular, the Iowa Farm Bureau championed the concept put forward by the Reagan administration that the free market and the revival of exports were the best medicine for Iowa agriculture. In September of the election year 1984 the president of the Iowa Farm Bureau, Dean Kleckner (in 1986 he was elected president of the Farm Bureau Federation), and Dixon Terry, a leader of the coalition, debated farm policy at Waverly, Iowa. Their meeting provides a chance to sample the ideas of their respective organizations at a crucial stage in the farm crisis.

The smooth, articulate Kleckner, the hog farmer from Rudd, Iowa, who would be roundly booed at the Ames rally in February 1985, epitomized the Iowa agribusiness establishment. Dixon Terry, a bearded, thoughtful Adair County dairyman, who had dropped out of Iowa State University to live in a commune and work as a janitor and garbage collector, was a proponent of sustainable agriculture. Predictably, they disagreed on just about everything.

For the Farm Bureau, Kleckner maintained that government policies had hurt the ability of American farmers to compete in world grain markets. The government, he said, needed to lower support levels so more corn could be sold abroad. Terry disagreed. He attacked the take-over of agriculture and the food delivery system by giant corporations. By encouraging farmers to produce more, he said, government policy had been pushing small farmers off the land since World War II. The result was a concentration of resources in fewer and fewer hands and the loss of Iowa's "rural character and culture." Iowa's "family farming system was a way of life," something "very precious," and losing it would be a tragedy. Although Kleckner agreed that things were "not good in agriculture," he saw government economic policies, not corporations, as the cause. Government policies created inflation, and runaway inflation hurt farmers by encouraging them to expand, sometimes with a push from their lenders. The government caused further damage when it changed policies to curb inflation. Supply and demand, Kleckner urged, was the answer. As the debate ended, Kleckner, unperturbed as ever, somewhat patronizingly acknowledged the coalition as "responsible, respectable, people," despite their misplaced ideas.[20]

Over the next two years both sides would continue to express their views. The coalition would consistently campaign for acreage controls and higher support programs as cornerstones of the 1985 farm bill. The Farm Bureau, as spokesman for agribusiness, saw crop-reduction schemes as harmful to the grain trade and the farm service industry and would continue to advocate the free-market responses of supply and demand.[21]

In the early stages of the farm crisis the coalition made progress that would later force mainstream organizations to play catch-up when it became obvious that the debt crisis would not go away. Early efforts, like the Farm Crisis Day at Nevada, Iowa, in October 1982 and the distribution of a research paper titled "The Crisis in Iowa Agriculture" in July 1982 were scorned at the time by more orthodox observers of the rural scene.[22] Like any cause in the public interest, the coalition found it essential to get favorable publicity and succeeded brilliantly at this task. They built good relations with the major newspaper, the Des Moines *Register*, whose coverage was vital to alert the public to the problem. In addition, the offices of the coalition in Des Moines became a way station for every national and foreign print and television journalist working on the farm problem in 1984-1985. For a short time in late 1984 and early 1985, when Iowa was reeling leaderless because the governor and the legislature could not decide what to do about the crisis, the coalition seemed to be filling the vacuum.[23] The moment passed, however, and the establishment finally overcame its inertia and began to act against the crisis.

It is interesting to consider how a small group of progressive activists, some of whom had no connection to Iowa or to agriculture, attained such a position. First, the cause they were fighting for was sacrosanct. Although some of their positions were controversial compared to those of the more conservative farm groups, because they were "fighting for family farms," their stand was unassailable. Unlike other activists, working for less popular causes, they avoided being smeared with the brush of "troublemaker," partly because they included church groups in the coalition and partly because they were able to acquire unique evidence of the growing crisis through their hotline.

For two years the office in Des Moines was alone in its monitoring of the condition of farm families at a personal level. Coalition counselors, unlike state agency personnel, were free to provide callers advice on legal, financial, and emotional matters. Usually, callers were in distress because their lender had begun to squeeze them. They needed financial advice and often legal advice as well. In such cases, counselors were able to furnish the names of attorneys with some competency in agricultural law. Sometimes family members called in a state of severe emotional stress, and counselors had to steer them toward another kind of help. Quite soon an aggregate picture of the state was built up, as certain lenders initiated policies that required rapid action on the part of borrowers.

Grass-roots organizing was also part of the coalition's basic program. They began in the southern and southwestern counties, where crops had failed. Through public meetings, the coalition recruited advocates to deal with local problems, and these community crisis committees began to be active in the counties within easy reach of Des Moines. In the winter, especially during the crucial period when farmers sought funding for the coming year, the coalition organized workshops in many localities both to ignite community interest and to educate farmers on the issues. The workshops also served as a training ground for advocates who would be counseling on legal, financial, and emotional problems. These sessions usually consisted of day-long workshops, often followed by a general farm crisis meeting, open to the public, at night. The training sessions focused on what a family needed to know when faced with accelerated foreclosure by a lender—the appeal procedure under Farmer's Home Administration regulations and the pitfalls of a bankruptcy proceeding, for example. Similar kinds of programs, organized by Catholic Rural Life, often took the form of religious retreats. These sessions stressed a more holistic approach to change in agriculture, discussing sustainable agriculture and the implications of the take-over of the state by corporation agriculture, "the El Salvadorization of the countryside." Such concepts were not entirely familiar to an audience used to an agribusiness view of farming.[24]

Fortunately for farmers, the Iowa Farm Unity Coalition was in a position to provide leadership at a time of extreme uncertainty. By and large, inflation had further sapped the infrastructure of rural society, weakening strong rural neighborhood ties. Farm families in Iowa were always the most independent segment of society—traditionally planning their own financial arrangements within the family or at most the local community. Suddenly, a substantial segment of the population needed spiritual and psychological counseling and financial and legal help that few were trained to provide. Not many Iowa lawyers, for instance, had had much experience with bankruptcy law as it applied to agriculture, nor were the "helping professions," the ministry and social workers, prepared to deal with farmers. During the sixties and seventies, even rural churches had concentrated on issues of more concern to urban or suburban congregations. Seminary training gave scant attention to the rather specialized ministry to farm families. Likewise, psychological social workers had little or no experience with farmers. By 1985 these professions were rushing to fill the gap.[25]

Probably the most important contributions of the coalition during the premobilization phase were their proselytizing for an alternative system of agriculture and their advocacy in the areas of farm finance. The diffusion of basic knowledge about the rights of borrowers and about foreclosure, replevin, and bankruptcy gave farmers some of the tools necessary to meet lenders on more equal terms. During the mobilization phase, the coalition would concentrate on other activities—lobbying both at the legislature in Des Moines and in Washington; the initiation of a bank-closing team to inform borrowers in a failed bank of their rights under the Federal Deposit Insurance Corporation rules; counseling on class-action suits against lenders; and a continuation of advocacy workshops, especially for professional groups.

Families in Crisis

It might be assumed that those farmers who emphasized familism, which tended to generate a long-term commitment to farming and frugal, conservative methods of operating, would have a better chance in the downturn than others. A tradition of working together meant that some effort was made to ensure that any major decision benefited the whole family. At the same time, overall authority remained with the senior generation, whose instincts rested on the side of caution, though conservatism could also alienate younger family members who wanted to move at a faster pace. But the influence of familism was diluted by the complexity of modern inheritance and estate tax law and the increasing emphasis on the farm operation as a business.

The generalization that farmers are good at raising crops and livestock but less proficient at managing their operations has a ring of truth for all farm families, regardless of their background. Farm finance and agricultural law are esoteric disciplines over which few have command. Moreover, the havoc created by a declining economy was a new and unpleasant experience even for the experts.

By and large, in the years 1982-1984, many families had to deal with economic difficulties alone or with minimal assistance from relatives, neighbors, friends, and professionals. The stigma of failure was still powerful, and those who did have problems were looked upon locally as "poor managers" and blamed for their own difficulties. Moreover, lenders who would later be restrained by court action, public opinion, and the realization that such an approach was not productive, were inclined to pursue liquidations aggressively because they themselves were under pressure from bank examiners to improve their collateral position. All over the state lenders adopted procedures to retrieve assets from farmers who seemed to be showing signs of failure or poor performance. Lenders would ask farmers to sell land or income-producing assets like machinery or feeder pigs; would demand that other relatives cosign notes or mortgages; would shorten the maturity date on notes, increase interest rates, refuse to release funds for living expenses, threaten liquidation. Some even used telephone harassment in the dead of night.[26]

The period of confrontation has been described as a time of psychological violence against farm families. It was violence inflicted on paper through foreclosure notices, farm sale advertisements, legal briefs, and court orders. Slowly, farmers were economically strangled by conditions beyond their control, which they did not understand. It is not surprising that some farmers, "insomniac, pacing, and impatient for explanation, considered the actions by lenders as a conspiracy against them, and turned to false prophets who offered appealing answers as to whom should be blamed—the Federal Reserve, the trilateral commission, international bankers, the Jews."[27]

Horror stories abounded. Farmers were tricked into signing over property free and clear to a bank; grandparents were asked to surrender time deposits as collateral and then grandchildren were foreclosed on anyway; oral agreements were reneged on; crops were harvested for farmers, and then the farmers' shares were withheld. A family that farmed 428 acres in Delaware County had their machinery and cows repossessed by a finance company after the family had filed for bankruptcy, when all legal action against them should theoretically have ceased. The creditors simply arrived one day and loaded the cows into a truck contaminated with hog manure. Later, the finance company tried to repossess fifty cattle belonging to a son, claiming they were registered in his name so that the parents could avoid losing them in foreclosure. Another farmer discovered that his bank had taken an inventory of his property in his absence a week after he had filed bankruptcy. Considerable friction developed when Iowa passed a law in the spring of 1985 that forced buyers of farm products to write checks out with both the farmer's and the lender's names on them if the purchased products were signed over to the lender as collateral. Using this lever, once checks were deposited, some lenders refused to give families any money to continue doing business. One bank refused to give a farmer enough money to purchase medicine for his epileptic child.[28] Another bank in northeastern Iowa pursued liquidations so aggressively that it contributed to one farmer's suicide and resulted in a number of well-publicized foreclosures that poisoned the fragile relationship between this lender and borrowers in the community.

In such a situation mistrust came to characterize relations, and a "we-they" syndrome developed. It was a learning period, when farmers and lenders made costly mistakes. For example, one suicide in the summer of 1984 might have been prevented, but the bank showed poor judgment when it cut off operating and living expenses to the family; a lawyer hired by the family had little knowledge of bankruptcy law and the rights of his client; and family members failed to appreciate the vulnerability of their father to the extreme pressures a lender could generate.[29]

But farmers learned fast. Through conversations in the coffee shop, the tavern, the sale barn, and around the kitchen table, they mapped strategies for dealing with the new climate of confrontation and how to deal with it. Often they became better prepared than the very institutions to which they were indebted. A measure of their success can be drawn from the remarks of a banker who decried "attorneys, self styled 'consultants,' and others recommending that distressed borrowers become experts in evasion, that they sue the lender for alleged 'bad advice,' or demand a write down of their obligations."[30] It is impossible to judge the amount of "creative activity" employed by farmers in their efforts to stay afloat. It is sufficient to say that for every farm family who acted innocently and naívely about its financial affairs, there was one that used every device at its disposal to remain in farming.

One of the favorite instruments in 1983-1985 was the Chapter 11 bankruptcy—a very blunt and rather expensive tool designed not for a firm as small as a farm but for an industrial concern like the Milwaukee Road. Its one advantage was that it kept creditors at bay while a reorganization plan was drawn up that would permit the firm to stay in business while paying off debts. Its major drawback was that in a poor economy it was difficult to generate enough income to make a reorganization plan work. On the other hand, a Chapter 11 had a strong attraction for hard-pressed farmers. Once they were in Chapter 11, any additional property acquired remained in their possession, so that it was possible to generate income that was unsecured. Using hindsight, it is obvious that the Chapter 11 solution was far from ideal: it demanded large attorney fees up front, and over the long haul its success rate was tiny. But often in 1983-1985 the farm family had nothing to lose and nowhere else to turn. Many farmers were too old to choose an alternative career, even if jobs had been available. Moreover, there were the Catch 22 implications of income taxes and deficiency judgments. Not only did the Internal Revenue Service treat foreclosures as sales and expect capital gains taxes to be paid on them, but lenders could demand a deficiency judgment if the sale of the property did not meet the amount owed. A full bankruptcy under Chapter 7 would wipe out a deficiency judgment but not the taxes owed on property. Indeed, even if a farmer declared full bankruptcy before foreclosure took place, he was still obligated for certain tax liabilities.[31] Not until the spring of 1986, when state and federal legislation abolished the alternative minimum tax, was this hurdle eliminated for distressed debtors. Even now, however, the whole issue of tax liabilities remains blurred and will probably take some years to resolve.

The evidence seems to suggest that the flirtation with Chapter 11 bankruptcy among Iowa farm families was symptomatic of the evolutionary way in which the crisis developed: it was a solution for a time, but then it passed. Initially, the two bankruptcy judges in Iowa were sympathetic to a farmer strategy of using Chapter 11 to stave off creditors. Eventually, however, the volume of business made them

Table 16. Bankruptcies in the Southern District of Iowa

Year	Chapter 7	Chapter 11
1982	85	28
1983	132	39
1984	192	64
1985	379	48
1986*	347	34

Source: Bankruptcy Court, Southern District of Iowa, Des Moines.

*First eight months.

revise their earlier assessment. One judge virtually worked himself to death trying to keep up with the case load. The other, Richard Stageman, who became quite outspoken, saw farm bankruptcy as a more serious dislocation than those faced in other occupations. "The displaced farmer," he said, "is not a blue-collar worker who is laid off. He is not an executive let out. He is not a merchant fallen on hard times. . . . These latter unfortunates, suffer, but usually temporarily, in transit to like labor, trade, or business positions. The established landholding middle-aged farmer whose life began on the farm, whose family is part of it, who has never considered another life, faced with its loss feels that this is the end for him and his."[32]

A year and a half and hundreds of farm bankruptcies later, this same judge was distraught over the experience of trying to deal with a serious economic and social problem through the bankruptcy court. In an emotional response to questioning at a federal hearing called to investigate a possible change in the bankruptcy law for farmers, he expostulated that the bankruptcy court, especially Chapter 11, was not the way to deal with the farm crisis. It would be more appropriate for the government to take the whole of the southern two tiers of Iowa countries out of production, than to expect results from a system designed for a mercantile debtor. Debt-ridden farmers simply could not face reality, could not abandon the hope that someone or something would come to their aid. "This congenital optimism is really a pipe dream," he remonstrated. Stageman estimated that he had heard more testimony from farmers, country bankers, and farm suppliers than any judge in history. "I cannot count the instances," he said, "when I have witnessed adult, self-contained, weathered men give testimony punctuated with despairing sobs. There have been too many nights that I have laid awake at night with tears in my own eyes reflecting on the day's proceedings." Efforts to "preserve the myth of American Gothic" by trying to protect farmers from harsh economic realities were "hopeless." This pessimistic appraisal was borne out by the statistics. In the northern district of Iowa, there were 87 Chapter 11 farm bankruptcies filed in 1983, 198 in 1984, and 228 in 1985, but because of the continued downturn of the economy only a handful of these cases were ever confirmed. After a futile period of struggle many converted to Chapter 7.[33] And the situation was no different in the southern district, where figures are more complete (Table 16).

Straight bankruptcy, or Chapter 7, which required complete liquidation of a farm business, was used once all other avenues for resuscitation were exhausted. Regardless of the instrument used, the rush to bankruptcy was remarkable in view of the traditional stigma, the expense, and the social implications. Typically, farm families had spent several generations in a given community, and local relations, friendships, and obligations often made the decision to seek bankruptcy most difficult. The testimony of one farmer from northwest Iowa, a college graduate who had joined his father on the farm in the sixties, reveals the agony such a decision entailed.

Despite hard work and the expenditure of large amounts of borrowed capital, one undertaking after another found itself in trouble by the middle of the seventies. First cattle feeding, then the elaborate farrow-to-finish hog farm failed to make the kind of money necessary to sustain a father-and-son partnership. To make matters worse, this particular son was buying his father's share, but buying a share did no more than shift the debt from one balance sheet to another. Since the farm was not earning enough to pay for itself, the operation became increasingly burdened with debt. The final straw came when a projected price for hogs did not materialize, and yields for both corn and soybeans were disappointing. Strapped for cash, the son knew that a Chapter 7 bankruptcy was the only way he and his father could make a clean break from their indebtedness, but there were all kinds of pressures against such a clear-cut solution. He was president of the local school board and prominent in other civic groups, and he feared that in losing the farm he would also lose the respect of a host of friends and neighbors. His own words best describe his final decision:

> How could I let this happen? The sum total of Dad's lifetime of farm earnings and all his net worth which we had accumulated in 17 years of farming—gone. No matter how frantically we tried to generate cash flow and cut expenses . . . each year our debt would creep up a bit. . . . Maybe it is time to admit that the present structure is not working and try to start over.
>
> Since Dad's net worth is gone because of my decisions, . . . would not we both be better off to do whatever is necessary to make the quickest recovery possible? What about my wife and children? What would have to occur in our present operation to be able to help with college expenses in only four years? In our present situation, the recovery would need to be spectacular. . . .

In early December, I spent a great deal of time contemplating the moral and ethical considerations of walking away from debts that I had agreed to pay. How will I be able to maintain my self-respect in the presence of my creditors who lost money on my account? What attitude will other members of the community have toward my father and me? My family has always stressed the importance of honesty in all dealings with other people. We recognize that once a person's honesty has been marred, it can never be fully restored.[34]

Once the son made his decision to file bankruptcy, the anxiety over ethical questions abated, but it returned to haunt him in an emotionally draining period between the filing and the actual hearing. Creditors began to harass him and his father unmercifully, and one sued for a twelve-thousand-dollar debt. After several months of misery, the hearing was held. Surprisingly it turned out to be something of a letdown. With so many hearings scheduled on the same day, it lasted only a few minutes. Instead of the traumatic grilling by creditors, which the family had dreaded and prepared for, there were only a few questions and answers in lowered voices. In May 1985, the family planted another crop of corn and soybeans, thanks to the generosity of their landlord, who waived a deficiency payment. The family had managed to remain in farming for one more year, but at considerable psychological cost.[35]

With backs up against the wall, families began to cross the fine line between legal actions and illegal "creative activity." The selling of sealed government corn or the removal of livestock and machinery before a scheduled repossession were just some of the ways desperation got the better of good judgment. Similarly, lack of alternatives, poor information, ignorance, and the need to blame someone channeled some into far-right groups. These shady organizations never had any kind of stable membership, but their views were attractive to desperate people.[36]

Many ultraconservative groups believed that conspiratorial forces were at work in the world, that Jews controlled the Federal Reserve System and the world economy, that citizens should arm to protect themselves, and that gold and silver were the only legal tender.[37] Farmers who held these kind of beliefs resorted to do-it-yourself lawsuits against local county officials involved in repossessing property. Some hard-pressed farmers struck back by filing documents designed to bog down the court system and to harass public officials and lenders. According to one farmer who used these methods, lawyers, judges, and lenders, were "all working together to gain ownership of farmland and sell out farmer after farmer." Do-it-yourself materials included legal karate packages distributed by a right-wing organization in Wisconsin. These contained instruction on how to keep possession of property, how to sue banks for breach of contract, fraud, and racketeering, and how to educate the local sheriff on the credit issue.[38] Fundamentally wrongheaded, such tactics distracted farmers from developing constructive responses to the problem. Most steered clear of this kind of activism, however, and it was exposed in its true colors by the news media and legitimate farm advocacy groups.

There were a number of symbolic protests aimed at spreading the message through media attention. The Iowa Farm Unity Coalition used this tactic in 1984-1985, sponsoring rallies designed to forestall and protest foreclosures. These had the effect of forcing the Farmer's Home Administration and the Land Bank to back off or review its procedures in cases where there were extenuating circumstances to a farmer's failure to make payments on a loan.[39]

Even more significant was a series of carefully chosen lawsuits designed to test the fiduciary relationship between farmer and lender—in other words, to try to show the dual responsibility between the lender and client in cases involving large losses. In one such case, a farmer sued the Southeast Iowa Production Credit Association for agreeing to finance an expansion in the seventies and then reneging on a promise to renew the loan. In this instance the jury not only canceled the debt but also awarded $1.3 million in damages to the farmer. Since juries in rural areas were usually composed of farmers and their relatives, it is not surprising that the jury trial strategy held promise.[40]

Often farmers found that bureaucrats in FmHA offices were antagonistic to their needs or slow to comply with new favorable guidelines sent down to the county level. On one occasion in northeast Iowa, activists arranged for a visit by the state director to hear grievances of farmers who had failed to obtain emergency and regular loans. With television and print reporters present, the state director was sufficiently impressed by this show of grass-roots resolution to insist that the local FmHA office begin to improve its performance and delivery of service.[41]

By the time farmers had begun to think about operating money for 1986, the farm crisis had entered a new phase. The debt crisis finally began to be resolved. Creditors, especially those holding private contracts, realized the futility of trying to sell land in a depressed market in which land prices continued to decline. Like their predecessors in the Depression, they began to see the wisdom of renegotiating contracts. It was better to lower interest rates and even principal to more realistic levels than to face the expense and frustration of foreclosure.[42]

Farm families, then, remained in the center of the firestorm that was the Iowa farm crisis. Three years passed from its earliest days, when few recognized its existence and victims had to make difficult decisions without help from outside, until the state mobilized to provide a veritable cornucopia of assistance. The relative slowness with which the momentum developed underlines a particularly unpleasant aspect of an economic crisis in agriculture—that the seasonal nature of farming kept so many families in limbo about their future for long periods. Waiting to see whether they would get crop financing, waiting for the foreclosure notice, and waiting for the machinery sale only prolonged the intense psychological stress. The large numbers of families in difficulties at one time did not simplify matters. Competition for FmHA loans at reasonable rates with guarantees and for off-farm jobs at minimum wage was fierce. And all of it came on the heels of one of the more prosperous and hopeful times in Iowa agriculture.

Lenders

Like farmers, lenders had no experience in dealing with an economy in severe recession. In encouraging expansion and leveraging in the seventies, they had acted no differently from their urban peers, who pushed clients into questionable schemes in Latin America and in the oil fields of Oklahoma.[43] The boom had reinforced the old ways of doing business in country banks. Handshakes, gentlemen's agreements, and friendly relations characterized the seventies, and bankers were often lax about evaluating their customers. Even if returns were poor, inflation in land prices would presumably cover bad management decisions.

When hard times hit there had to be a revolution in such business practices. There was a rapid turnover in personnel. Suddenly, tax records were demanded, and banks resumed frequent customer evaluation. They began visiting farms to check inventory and started requiring monthly financial statements. Farmers found this constant scrutiny galling. Formerly friendly business relationships were contaminated by suspicion and even paranoia, which made communication difficult. Whereas in the boom times, animosity would have been worked out through the migration of accounts seeking better terms elsewhere, under the new climate such flexibility, particularly for poor-risk accounts, was impossible. As bankers pointed out, these changes in client relations were as much the doing of the farmer as the lender.

The climate of increased polarization and potential for confrontation was well summarized in testimony before the House Banking Committee in the fall of 1985. The changing attitudes in rural America had at least three dimensions:

> First, business relationships . . . are deteriorating. Farmers who were willing to cooperate with their lender in making adjustments in prior years are taking a more protective-of-self stance. Merchants and dealers are less willing to operate without excessive legal documentation of transactions. Business

> people are becoming more suspicious and less trusting in their dealings. This "non-cooperative" attitude shows up clearly in the lending relationships where farmers use the threat of bankruptcy to gain accommodations from the lender. Some farmers are "building new houses," separating real estate and other assets from the farm business and their own personal ownership by transfers to children . . . to protect property from creditors and have a base to restart if the "old house"—the current farm—is lost. Second, some farmers who have the financial ability to pay . . . are consciously debating whether they should do so. This is particularly the case with the Farm Credit System's borrowers where the discussions of financial assistance or a "bail out" are most frequent. . . .
>
> There appear to be changing standards of "honesty" or "commitment" in rural communities compared to earlier years. The "your word is your bond" attitude is no longer standard. Rural people are not necessarily becoming blatantly dishonest, but they are more willing to accept the grey area between "right" and "wrong." . . . The reasons for this change are twofold: one, that people's standards sometimes are adjusted when financial survival is at stake, and two, farmers feel that their current financial problems are "not their own fault" and that others—including lenders, business firms, and the government—are partly to blame.[44]

During the period defined here as confrontation, there was considerable mistrust between bankers and their clients. At one community farm crisis meeting in northwest Iowa in early 1985, the proceedings developed into a shouting match between a priest and a banker. At another meeting between bankers and activists, it was obvious that several of the bankers were extremely upset by the perception that they were being singled out as the rogues of the local community. When they later walked into a public meeting in a group and sat down together, a murmur of disapproval went up through the assembled crowd. To their credit, these men, and many like them, were prepared to face hostile audiences in community forums, to listen to verbal abuse over the telephone and in their offices, and to wonder whether they and their families would suffer more serious harm.[45] Lenders complained of the extremely stressful conditions in which they worked. According to one report, 59 percent of rural Iowa bankers claimed to have been verbally abused in 1985.[46] To many, it seemed that the whole world was ranked against them. Politicians, ministers, even their own family members questioned their role as liquidator of the family farm.

The pressure on country bankers came not only from their customers but from bank regulators, who began to scrutinize problem loans in 1984-1985. The debacles suffered by Penn Square and Continental Illinois increased the vigilance of regulators and the paranoia of bankers. Indeed, in testimony before congressional committees, Iowa bankers accused the regulators of employing a double standard when reviewing foreign loans in large banks and farm loans in country banks. According to the testimony, it was the harsh standards applied in rural areas that forced the liquidation of many farmers who might have stayed in business had the bank regulators not turned the screws.[47]

The closing of a country bank had a special impact on farm borrowers. Almost all the banks that failed in Iowa were insured by the Federal Deposit Insurance Corporation, and failed banks typically reopened the following day under new management. However, while the process usually worked smoothly for depositors, for borrowers—most of whom were farmers—the transition was often long-drawn-out and frustrating. The new ownership had an exclusive thirty-day right to purchase the loans of the failed bank. After the thirty-day period, other institutions were also allowed to purchase loans. Then, the FDIC liquidated the remaining loans not purchased by the new institution or by others. In Woodbine, Iowa, where a bank failed in February 1985, only 25 percent of the agricultural loans in the old bank's portfolio were picked up. For the others there ensued a period of frustrating uncertainty while the borrowers, the FDIC, and the new institution attempted to find alternative funding through the FmHA.[48]

It would be incorrect to portray lenders as showing a united front against an aroused farm community. Considerable confusion existed just before and during the planting season in 1985, when the Farmer's Home debt adjustment program, was being offered to banks and the Farm Credit System. According to one banker, attempting to work with the bureaucracy of the FmHA was a nightmare. Most bankers wanted more than anything "to have these difficult farm problems solved and have worked very hard to get these applications processed. But we have been frustrated by red tape, technical interpretations, and delays; and the amount of paperwork is ludicrous."[49] Another banker was less inclined to blame the bureaucrats. "Let's not shoot the messenger," he warned. In his opinion the trouble came from higher up, in the Reagan administration, which he had voted for in the last election.[50]

Both the FmHA and later the Farm Credit System, which in the summer of 1985 came under congressional scrutiny because of its own losses, were favorite whipping boys of conservative Republican politicians in Iowa who were trying to distance themselves from the Reagan administration. The senior senator from the state, Charles Grassley, chaired a number of meetings in 1985 to study FmHA and farm credit procedures. At a ten-hour hearing at Des Moines in September, Grassley and some Iowa congressmen grilled farm credit officials over their failure to recognize the farm crisis sooner and their lack of compassion for farmers in trouble. In reply to a spirited defense of the Farm Credit System, Grassley became uncharacteristically agitated:

"Listen," he said, "we are all in this boat together, I see a lot of people around here that have devoted the last year and a half to just keeping farmers from committing suicide. We are in a situation where we aren't going to follow the usual economic laws to try to pursue an answer. We are talking about a problem of humanitarianism. We all have to be involved in this business. We are going to have to work together to find solutions."[51]

Like the banks, the Land Bank and the Production Credit Association acted unwisely when confronted with failing customers. One Iowa congressman conducted four town meetings in his district in the summer of 1985 on the credit situation, and the message he received about the Farm Credit System was disturbing. There was no forbearance; farm member control was being usurped by hired staff; no information was released to members; people were foreclosed because one payment was late; and alternative financing was not considered. Management apparently refused to accept any responsibility for the condition of the system. Instead, they pointed to the general farm situation and cried, "Foreclose, foreclose."[52] By 1985 many Federal Land Banks and PCAs were, in the words of one director, "floundering financial institutions on the razor's edge of liquidity." Morale was low, and turnover was high.[53] "One employee made an agreement, and then he was moved on," said a knowledgeable insider about the system. "Another employee comes in and denies that that verbal agreement was ever there. . . . So we have a large volume of structural—what I call structural—lies inside the institutions. It is not planned. It's just that the hired help has not got the ability to do this. Most of the employees that are dealing with 65-year-old men, telling them to go out of business are in their late twenties."[54]

Not surprisingly, the Farm Credit System spent much of the final quarter of 1985 getting its own house in order and merging Land Banks with PCAs. It issued a moratorium on foreclosures during the final months of the year when 960 Land Bank and 170 PCA foreclosures were pending in the Omaha district. Officials were extremely cautious in their predictions of foreclosure activity once the moratorium was lifted. Apparently they planned to resume, but "only for those borrowers where the association has reached a conclusion that there are no alternatives."[55] It is worth noting that once mobilization was under way, the foreclosure issue was treated with kid gloves. The governor of Iowa signed a foreclosure moratorium into law in October 1985—a more or less symbolic gesture, made under heavy pressure from farm advocates. Even before, however, lenders were working under certain constraints. For example, the FmHA, which had by far the largest number of problem loans, was barred from action because of a federal judge's ban on foreclosures. The judge also ordered the agency to draw up new rules to clarify the rights of borrowers. Even after the rules were published, there was a delay in implementation, and only those loans delinquent for over three years were acted upon.[56]

Paradoxically, the Hills murder case helped both sides begin to repair the rift. The incident, in which a distraught farmer killed his banker, a neighbor, his wife, and then himself, ironically, involved an institution with only a small farm portfolio, a farmer only marginally in trouble, and a banker with a reputation for fair dealing. As a symbol of how far things had gone, the murders sent a powerful message to the industry. Both lenders and farmers started to reevaluate their positions, lower the level of antagonism, and compromise on problems that separated them.[57]

Thus, they entered the 1986 credit season far more experienced at handling the psychological baggage that accompanies the delicate relationship between lender and borrower in a time of economic stress. By the spring of 1986, lenders had initiated what might be termed "social lending policies." Many loan officers had received training in the human dynamics of lending; they had attended role-playing workshops and had learned to guide farmers through the process of change. With the assistance of mediators, lines of communication were opened with borrowers, and the best alternative among unpalatable solutions was found for each case. Often such individual attention meant extra paperwork, and working with the Farmer's Home Administration for a guaranteed loan might require many, many hours to complete. However, this new effort apparently did not apply to those farmers who had already made an attempt to save their economic position through court action, and though some PCAs managed to work with those who needed forbearance,[58] the Land Bank gave few concessions, gaining a reputation for utter ruthlessness. A sample of Chapter 11 cases from the bankruptcy court of the Northern District of Iowa in September 1986 showed that both the PCA and Land Bank voted to reject reorganization plans in virtually every case. Presumably the huge losses suffered by the system can be blamed for this extreme inflexibility.

Obviously the farm crisis in Iowa had made as much impact on lenders as it had on farm families. Economic stress affected small commercial banks first, and twelve closed in 1985. Later, the federal credit system and the larger chain banking corporations came under severe strain. Because of the federal system's size; the red tape involved in its operations; its poor management, both locally and at headquarters; and the unpopular forced reorganizations of its subdistricts, it was unclear if it could ever resolve its economic problems. Many commercial banks, though often in serious difficulties, had one advantage over the Farm Credit System: they were not restricted to agricultural lending. In general, during the year ending in May 1986, the delicate relationship between farmer and lender had been rescued from the depths. Mediation, rather than vilification and confrontation, had become the rule.

The Establishment

For our purposes, we might consider the farm establishment as comprising such institutions directly involved in agriculture as the Cooperative Extension Service, the Farm Bureau, and the specialist commodity organizations—pork producers and corn growers, for example.[59] The establishment was not involved in the early stages of the crisis. Only later, when some found their credibility damaged, did they move to improve their image. For this reason, the establishment's involvement need only be analyzed during the mobilization phase, when these agencies made their chief contribution.

Whereas the Iowa Farm Unity Coalition envisioned resolution of the farm crisis through a stronger rural America, with more farm families, less emphasis on industrialized agriculture, and invigorated communities—in other words, a reversal of the universal trends since 1945—the establishment saw the resolution of the crisis in a different light. While the Extension Service mobilized, as best it could, to give farm families immediate assistance, its commitment to production agriculture and its ties to agribusiness made it difficult for the agency to alter its positions appreciably. For this reason, much of the aid given by the Extension Service was designed to ease the passage of farm families off the land into other occupations and to part-time farming, rather than to offer an alternative blueprint for agriculture. Playing the role of advocate was not part of the Extension Service mandate. Like farmers themselves, the service was heavily dependent on federal money for its day-to-day existence. With the USDA under pressure to trim its budget, the Extension Service was in a difficult position and unable to provide leadership on the issue of change. Probably the low point for both the Farm Bureau and the Extension Service was the winter of 1985. In the depths of the downturn, farmers joked about the three enemies of Iowa agriculture—the banks, the Extension Service, and the Farm Bureau—but slowly the establishment organizations managed to rehabilitate their tarnished reputations.

The Extension Service's most visible contribution, its toll-free hotline, was originally funded through donations from the Farm Bureau, FmHA, and the Farm Credit System. Later large agribusiness firms such as Monsanto and cooperatives such as Land o' Lakes, as well as the state, contributed funding. In 1986 a special appropriation from the legislature, which included money for legal services and mediation, funded the hotline for another year.[60] The Iowa Rural Concern Hotline acted as a referral agency, putting farm families in touch with a particular service (legal, financial, emotional, educational) in the area where they lived. The nine counselors, including several ex-farmers, a lawyer, a Methodist minister, and a nurse, had all received special training. During March 1986, for example, while farmers were trying to complete financial arrangements for spring planting, an average of seventy-six calls were taken every day. Statistically, 67 percent of callers were male and 33 percent female. Over 90 percent of the calls dealt with financial questions. As many as 63 percent dealt with lender problems, and 66 percent with legal difficulties, of which about half were referred to the staff attorney. During the month there was an increase in callers discussing family problems; 10 percent mentioned problems with divorce, chronic illness, or alcoholism. About 9 percent of callers needed help with basic human needs, information about local food pantries, food stamps, and county health programs. Interestingly, only 6 percent requested information about retraining or educational programs. Finally, 24 percent of the callers in March 1986 discussed emotional stress with counselors. Out of these 32 percent were defined as mild cases, 47 percent were moderate, and 21 percent severe. From the time the hotline was established, in February 1985, through May 1986, 15,544 calls were logged.[61]

Although the data seems to suggest that farmers were only marginally interested in retraining, the Extension Service and especially its ally, the Farm Bureau, took what might be termed a management approach to the problem of disrupted farm careers. They stressed leaving the land as smoothly and gracefully as circumstances would permit, trying to sell the notion that quitting farming was a career change in the conventional sense, rather than the end of a way of life. The Farm Bureau quite outspokenly urged farm families to face the cold realities of farming in the eighties. A free-market approach and low commodity prices made grain farming and, in certain instances, livestock production part-time occupations at best. As far as the establishment was concerned, the days of farming as a full-time career were over, and farm families must resign themselves to the need for off-farm jobs. Accelerated structural change would result in a large number of part-time family farms interspersed with bigger farms run by closely held family corporations and those belonging to absentee investors run by managers.[62]

At the practical level, the Extension Service placed most of its emphasis where it had most experience—on financial assistance. Using a computer program based on a what-if approach, farmers were encouraged to look at their operations in a variety of ways in an effort to cut costs and improve the organization of their businesses. Other innovations included the use of statewide satellite broadcasting for crisis programming on a mass scale. But because it was prohibited from advocacy, the Extension Service was limited in what it could achieve in the public arena. Nor could it offer much to those under economic and emotional stress. Indeed, with the Farm Bureau, the Extension Service continued to sponsor Dale Carnegie–type positive-thinking workshops, which seemed not only insulting but a cruel joke to the farm family in severe economic trouble.[63]

Thus, the Extension Service was placed in a difficult position because of the farm crisis. By the spring of 1986, it had regained some of the esteem it had lost. As will be seen, compared to the performance of the California Extension Service, it was a very model of progressivism. But like the rest of the agricultural establishment it had much to live down after the uncomfortable winter of 1984-1985.

The Legal Profession

The role of the legal profession in the farm crisis was also controversial. Although there is no systematic data available through which to assess the performance of attorneys, the general impression is that lawyers did not distinguish themselves. Because farm bankruptcies were rare before the eighties, only a few lawyers in Iowa had had experience with them. The only contact most lawyers had had with farm families was for estate planning, and the profession had become reasonably proficient at saving their clients from the burdens of estate and inheritance taxes. In the downturn, taxes also played a large part in strategies to save a family farm, but few lawyers or accountants had dealt with that kind of taxation before.

For these reasons, farmers were at first poorly served by their counsel. In small towns with few lawyers there were often questions of conflict of interest when a lawyer representing a farmer also had ties with the local bank. Advocacy groups responded by creating a statewide referral service, which directed farmers in trouble to lawyers who had some experience or success with bankruptcy. Most of these lawyers were in the larger cities in the state, and they did not hesitate to charge large fees for their time. In 1984 and the first six months of 1985, bankruptcy chasing became common, and some lawyers built up large case loads.[64] Like the bankruptcy judges who presided in court, they soon found it impossible to give their clients adequate service. Many clients were farmed out to interns. Farmers often had legitimate complaints about their treatment in bankruptcy, especially the high cost in comparison to the return. In the summer of 1986 a dozen clients of one bankruptcy attorney had their fees lowered for hourly work on their cases. The bankruptcy judge of the Northern District of Iowa granted a request from the Farm Credit System, which objected to the high rates because attorneys, not creditors, received first payment in any bankruptcy case.[65]

At the outset of the crisis, lawyers moved along a largely uncharted course. According to Fred Dumbaugh, one of the leading bankruptcy lawyers in the state, it was impossible to know in 1983 what course agriculture would take. The fact that commodity prices continued to drop for the next three years made havoc of reorganization plans that depended on stable economic conditions to allow a Chapter 11 to work.[66]

According to Dumbaugh, in the normal course of events, farmers had basically three options when faced with economic distress—to do nothing, to negotiate and attempt to mediate, or to file bankruptcy. The first would obviously lead to disaster for any farm family. By the time they awoke to the need for action, it would be too late to save anything from the hands of the creditors. Inaction, unhappily, was common enough in the early stages of the farm crisis, and families innocent of the law and their rights paid heavily. Ironically the "let's make a deal" scenario, which by 1986 was the approved strategy among lenders and farmers alike, was the route leading bankruptcy attorneys preferred from the first. But the banks refused to cooperate. The relationship between lenders and their clients in difficulties was still one-sided, for by tradition, the lender had total control over the farmer. It was the banker who drew up financial statements and suggested when and what to sell, while the farmer passively signed the security agreements, notes, and other papers without really understanding their content. Not surprisingly, given this sort of relationship, negotiations were never a feasible solution in the initial stages of the crisis. In 1983-1984, when lawyers called bankers to try to avoid litigation, the loan officer typically hung up the telephone.[67]

A Chapter 11 bankruptcy, then, despite all its disadvantages, became one method for attorneys and their farmer clients to stave off forced liquidation. Since this form of bankruptcy was extremely time consuming, requiring monthly reports and financial statements to the court, it was very expensive. The going rate for a Chapter 11 in Iowa in 1986 was at least fifteen thousand dollars, although this stiff fee could be lowered when legal assistants did some of the work. The keys to any successful Chapter 11 were fairness to creditors and the ability to produce a feasible plan for cash flow, so that the judge would be able to approve the plan. Unfortunately, the deterioration of the economy undermined the best-laid plans.

The objective in any Chapter 11 plan was to rewrite the debt to separate unsecured debt from secured. A plan was put together by breaking pieces of debt off and throwing the unsecured debt out. For example, in an operation with debts worth $500,000, of which $200,000 was unsecured, the objective would be to restructure the $300,000 of secured debt so that the farmer would pay a fixed interest rate of 11 percent on the principal over a thirty-year period. Similarly, on a hypothetical operating loan of $400,000 from the PCA, $100,000 of unsecured debt would be scrapped and the remainder would be packaged into a loan to be paid off in fifteen years at 11 percent interest. A private contract at 7 percent interest would not be restructured, because the terms were reasonable. The restructured operation, with all its debts consolidated, would then begin to pay off its secured creditors through a single payment. The unsecured creditors, on the other hand, would receive only token payment. They would be fortunate to obtain ten cents on the dollar, about what they would have received in a foreclosure.[68]

Running a farm business in Chapter 11 was not easy. There were innumerable restrictions imposed by the court, which for the individualistic farmer were a considerable burden. Not surprisingly, only a tiny percentage of farmers successfully worked their Chapter 11 bankruptcies through. Those who did, almost without exception, had assets outside the farm business that largely contributed to their success. In any analysis, most Chapter 11 filings in Iowa were useful only as a stalling mechanism to prevent the seizure of assets temporarily. In effect, they bought a little time before the farmer took a straight bankruptcy.[69]

Thus, after a period of encouraging farmers to utilize Chapter 11 as a legitimate means of remaining in business, its very poor record of success forced a reevaluation in the profession. By the summer of 1985, the Iowa Bar Association had mobilized sufficiently to run statewide seminars on bankruptcy and workouts, and by the fall the state itself had hired a mediator to teach others to mediate in workout situations. After thirty-six months or more of confrontation and litigation, "lets make a deal" had become the official policy of the state and the legal profession in agricultural workouts. However, for many unfortunate farm families, this change in climate came too late. By the spring of 1986 Chapter 7 farm bankruptcies had begun to dominate the bankruptcy dockets. Often debts far exceeded assets. Many owed hundreds of thousands of dollars to the Land Bank and the Farmer's Home Administration. In addition, unsecured debts to elevators, seed dealers, and other local businesses often exceeded fifty thousand dollars.

The introduction of the new Chapter 12 provisions designed especially for agriculture in late 1986 was a hopeful step. However, given the unhappy experience of farmers with bankruptcy, the chances of its success did not seem promising.

The Media

The role of the news media in the farm crisis has already been noted. The Iowa Farm Unity Coalition was especially adept at maximizing its effect and exploiting the impact of television on the general public. Through such events as cross plantings, tractorcades, and nonviolent protests at sales and foreclosures, the coalition sought to heighten public awareness.[70] Perhaps the Farm Unity Coalition's most important coup in this area was their advisory role for the film *Country*, whose pivotal scenes involved the relationship between lender and farm family. This film, shot in Iowa in the late fall of 1983, was a box office success, and coalition participation allowed Hollywood to anticipate the evolving farm crisis by several months.[71]

The imaginative use of both print and television media by farm advocates was a classic case of the cultivation of a conscience constituency. Essentially, farm interest groups in Iowa were engaged in a public relations "hearts and minds" campaign on behalf of the farmer, directed with great effect at the general public and, through the voters, at the ineffectual occupants of the Iowa statehouse—the governor and the legislature. Media coverage enhanced the successful political lobbying effort. In this instance, not only was a moratorium on farm foreclosures signed by the governor, who was under unyielding pressure from pickets outside his office, but more important, at the legislative session in the spring of 1986, some legislation passed over the protests of the state Bankers Association, dealing with homestead exemptions and deficiency judgments. The new laws somewhat mitigated the burden on farmers who liquidated their operations. Ironically, by the spring of 1986, the Iowa Farm Bureau Federation had made a 180-degree turn on this and other issues dealing with the economic plight of the farm business. The federation found itself supporting its old nemesis, the Iowa Farm Unity Coalition. Such was the power of mobilization.[72]

In Iowa the print media were dominated by the Des Moines *Register*—an important actor in the farm crisis by virtue of its tradition of rigorous coverage of farm news. Although the Gannett chain bought the paper in 1985, little noticeable change took place either in editorial policy or in feature stories on agriculture. Throughout 1985 and 1986, reporters contributed a steady stream of material that kept readers unusually well informed on the progress of the farm crisis. As a molder of opinion and a legitimate leader in the mobilization effort and the resolution of the crisis in Iowa, the paper, it is worth noting, was not averse to supporting electronic-media-oriented extravaganzas that aimed not only at giving direct support to farmers but also at promoting general awareness of the farm situation. To this end, the paper rented a train in September 1985 to transport farmers and agricultural leaders to

Champaign, Illinois, for the Farm Aid concert. It was an effort to provide both relaxation and interaction among some of the state's agriculturists.

Self-Help and the Helping Professions

Of all the actors in the farm crisis, it was perhaps the helping professions—the ministry and social and mental health workers—that encountered the most difficulty in reaching farm families. And though other outside parties also met closed doors in trying to help farm families, those trained to intervene in critical situations found the situation especially frustrating.[73]

Certainly the problem had something to do with the individualistic ethic of the farm family and how the male head perceived his role. Seeking help was considered a sign of weakness and entailed a loss of face—hence the prevalence of the "do nothing" syndrome, which had such dangerous economic and psychological implications. On the other hand, farm women, unless they were totally dominated by their husbands, often did not share these attitudes. Their daily round on the farm had given them the skills to interact with others for support. Moreover, farm women were usually more attuned to the danger signals of economic and psychological stress and hence were more open to intervention from the outside. Nevertheless, resistance to the idea of seeking help could be reduced only by strong leadership from within the ranks of farmers or by an outsider with special charisma and rapport with the farm families.

At the Extension Service, whose self-education vehicle had for years been the bulletin and the circular, it was natural that at least part of their effort in the area of self-help would be through self-study materials. Thus, by the fall of 1985 the Extension Service had prepared an eight-session course with a facilitator guide for each lesson. Sessions dealt with such topics as loss, blame, communication, the roles of support systems, managing stress, and family finances. Like most Extension Service publications, the course was a valuable resource for anyone involved in a farm support program. Originally, the Extension Service tried to involve the clergy in a recruitment program for male farm facilitators. Recruitment, however, was not successful unless an especially charismatic extension agent or minister was able to influence the outcome.[74]

At the same time, the goal of all those involved in farm support activities, whether they were from the Extension Service or Prairiefire, was to attempt to draw out grass-roots leadership from farm families themselves, to make them masters of their own destinies. Obviously, the best advocates for farm families came from within their ranks, and many of the "self-styled consultants" castigated by lenders for stirring up "creative activity" in the economic sphere were farmers who had gone through the wringer themselves. And the corps of mediators trained around the state in 1986 were initially chosen by virtue of their own personal experience with bankruptcy and negotiations with lenders. This policy was later changed.[75]

One successful example of intervention in the farm community came from the Cedar Rapids–based organization Farmers Helping Farmers. Perhaps the key to their success is suggested in their title, for the director had both a farm background and a degree in social work. Like Prairiefire, Farmers Helping Farmers had no connection to the establishment, was familiar with the problems of farm families, talked their language, and had a number of young, well-educated, and energetic people on the staff, who were capable of giving a good impression at the crucial first meeting of a self-help group. Whereas an organization like Prairiefire concentrated on training counselors and group leaders in the economic and legal dimensions of the farm crisis, Farmers Helping Farmers founded community resource and youth committees and trained facilitators to coordinate them. By the spring of 1986, one year after its founding, Farmers Helping Farmers, had youth groups in fifteen counties and fourteen adult community groups in eastern Iowa. They had initiated a training program for facilitators, who were screened before they were accepted into the program. Training involved day-long workshops to provide recruits with the skills necessary for farm family intervention.[76]

Effective intervention with farm families under stress required a special talent. Unless practitioners had a natural bent towards farmers, it was easy to grow discouraged after making an honest effort with little or no result. The Northwest Iowa Mental Health Center in Spencer was generally recognized as a leader in the development of services for farm families. Their rural response program, which covered nine counties, comprised support groups, peer listening and youth programs, and later VISTA volunteers. It was preventive in orientation, designed at its most basic level to steer rural communities through the period of economic and emotional stress for farmers and their families. Farm families, it was found, generally needed support, not therapy. Therefore, all members of a community were encouraged to help develop a supportive environment.[77] This was not an easy task in an atmosphere of blame and denial. Coping with the loss of a farm was painful, and farmers experienced anger, hostility, depression, self-pity, guilt, and helplessness. "The grief and pain following the loss of a farm," wrote one researcher, "is similar to the grief felt over the loss of a loved one. This grief follows certain predictable stages, and includes: shock, confusion, strong emotions, guilt, complete recognition of the loss, and recovery. The stages often overlap. Some may last only a few minutes—others, months. It takes time to heal, and pain can reoccur."[78]

The program had only one agenda—to help families get through a day at a time and "share the pain, the anger, and the caring that exists. . . . There were no lessons to teach people how to grieve. . . . rather the pain is shared so that the pain can be gotten through." The group leaders were trained to be understanding, empathetic facilitators. The caring between members allowed them to overcome isolation and gave them a sense that others were experiencing the same kind of difficulty. In a safe, "controlled" situation that allowed grief to be expressed, there was usually a reduction in stress. Anger was spent and violence avoided.[79]

The successful peer-listening program was begun for those who for some reason could not attend support groups. Peer listeners were community volunteers who, often times, had gone through the same experience; they were trained in listening skills, legal and financial issues, and "resource identification." Listeners could call on the resources of the center, pastors, and other members of the community to assist them in their work.[80]

The clergy, too, played an important role in the farm crisis. According to one southern Iowa minister, "The church was the right organization in the right place at the right time." It was anchored in the local community and had the symbols, the congregations, and the faith to speak to the situation. "We can talk about compassion, justice, hope and community," he went on, "and no one else has that as their valid property." The church was just now discovering what it had going for it: "these were times to be an authentic church."[81]

At the same time, some ministers were uncomfortable with activism. Aware of the political implications of being too far "out front" on the issue of farmer advocacy, they stressed conciliation. "I am very careful to try not to divide the community," said one minister. "Bankers and farmers are in this together." But in southern Iowa, where the economic crisis was especially severe, political infighting did not prevent the church from leading the way in forming networks for emergency aid. Food pantries and clothing outlets were common in 1985. Several denominations banded together to secure an eighty-thousand-dollar federal grant for ten VISTA volunteers to coordinate food gathering, employment, and handicraft activities in eight southern Iowa counties. In addition to this kind of activity, many members of the clergy spent countless hours in more conventional activities, such as sipping coffee and just visiting with their parishoners, listening, caring, and supporting.[82]

Conclusion

By the summer of 1986 the farm crisis had been in progress in Iowa for four years. All those involved in its outcome—farm families, agribusinessmen, advocates, lenders, social workers, extension agents, clergy—felt burned out by their experience. One of the most troubling aspects they had to contemplate was that final resolution of the crisis might still be years away. The specter of two decades of travail in the twenties and thirties was not encouraging.

The symbol of the family farm provided a powerful driving force for an unprecedented mobilization of state and private organizations to assist beleaguered farmers and their families. However, even after four years, it was difficult to gauge the depth and impact of the downturn on many operations. Reliable data was scarce, and government programs and "social lending" practices (as the bankers preferred to call them) tended to prolong marginal farm careers. Undoubtedly, without farm advocacy, there would have been a more clear-cut resolution to the economic downturn in Iowa. What some termed the congenital optimism of farmers also caused many to pursue paths that in hindsight ought not have been taken. Yet it is difficult to blame them. Their experience was overwhelming and often so debilitating that it caused many Iowa farmers and their families to move in directions that later proved impractical.

Advocacy had a number of implications for those involved in the farm crisis. Mobilization had benefits for small communities, for it forced residents to work together to resolve specific problems, recreating a solidarity that in boom times had often been allowed to lapse. In the same vein the crisis brought the helping professions—social workers and the clergy—into areas they had previously ignored. Yet advocacy, though it gave farmers a temporary stay against farm failure, also introduced another set of problems that an abrupt departure to get on with another career would have avoided. To be sure, the uncertainties of deficiency judgments and battles with the Internal Revenue Service made an already complex decision to fight rather than rollover even more problematic. The sorry record and the high cost of bankruptcy showed that farmers often found themselves in a Catch 22 situation when it came to reorganization.

Because of the huge amount of unfinished business, resolution may prove the most difficult phase of the farm crisis in Iowa. It could turn out to be an extremely lengthy period, costly in both financial and psychological resources. However, as in the confrontation and mobilization phases of the crisis in Iowa, advocates for change in agriculture and rural communities could hold the key to eventual resolution.

In the Central Valley, quite different traditions in farm organization and in and how growers faced economic problems ensured that the downturn would develop along different lines. It is to this environment that we now turn.

9 Central Valley Crisis

The farm crisis slowly gathered momentum, and eventually it affected all farm regions, including the Central Valley. Although California agriculturists often preferred to use the term "deep recession" and usually made a point of emphasizing the state's exceptionalism—the diversity of crops and the more balanced economy in the Central Valley—essentially a similar scenario of doom and gloom had developed there by the end of 1985.[1] In addition, early in 1985 high concentrations of selenium were discovered in irrigation drainage water of the Westlands Water District. This major ecological threat would pose a serious challenge to the long-term health of the agricultural economy in the valley.[2]

Farm families faced the same situation as their compatriots in the Midwest—depressed prices for commodities, which translated into poor cash flow; difficulty in servicing debt; lower incomes; and most important, drastically lower assets because of plummeting land values. The farms facing the greatest problems were those that had expanded operations during the late seventies, going further into debt in order to acquire more real estate. The increased demand for agricultural land both from farmers as well as investors, who saw real estate as an attractive investment and tax shelter, caused a boom in California farmland. From 1978 to 1981, values increased at an average annual rate of 23 percent. As was true elsewhere, the higher prices could not be justified by a strict analysis of the income the land was expected to produce. As soon as agricultural real estate stopped appreciating, a familiar chain of events was set in motion. Cash flow replaced equity as the lender's criterion for financing operations. The highly leveraged, who were the first to feel the pressure, had difficulty in meeting obligations as early as 1984. By the financing season of 1985, the overextended were under fire from their lenders to sell off assets, and foreclosures and bankruptcies multiplied.[3] In the spring of 1986, only the self-financed were free of the extremely stringent rules set in place by lenders undergoing economic woes themselves.

The agricultural downturn in the Central Valley had a different rhythm than that in Iowa. Given the greater complexity of the agricultural structure, the diversity of the industry, the concentration of private agricultural loans in huge institutions such as the Bank of America, the different traditions of the Cooperative Extension Service, and the rigid dichotomy between labor and ranchers, variations both in the parties involved and the chronology of events are to be expected. In broad terms though, a similar model of denial, confrontation, and mobilization provides a useful framework through which to understand the events of 1984-1986. In a climate dominated by corporate agribusiness, which tended to deny unpleasant news or to present it in the most favorable light possible, there was a natural inclination to play down the many indicators of trouble and to assert that California agriculture was immune to the ills of farming in the Midwest. Nevertheless, the crisis did not take California farmers unawares. Unlike midwesterners, who had to confront an entirely new and unexpected situation with few available resources, Californians were able to profit from the experience of others. The metaphor of the prairie fire expresses the spreading nature of the crisis very well. In 1984-1985 many of the indicators of California's downturn were unclear, but the local media gave the farm crisis in the Midwest ample coverage. Farmers in the Fresno viewing area, for example, could turn on their television for the morning farm show and see programs that originated in the Midwest. They became familiar with all the symptoms of crisis and were capable of recognizing them in their immediate neighborhood, but not surprisingly, at least at first, they preferred to ignore them.[4] Ranchers interviewed in October 1985 still hung on to the illusion that the state would escape relatively unscathed. By contrast, farm suppliers, gins, and giant commercial lenders had already passed through the denial phase by then and were mobilizing their resources to avoid the shocks a confrontational atmosphere would bring.[5] By the spring of 1986 farmers and ranchers felt the changed atmosphere of their lenders' offices and suddenly experienced what some of their peers, who had showed early symptoms of economic stress, had been through in recent years. In short, momentum had begun to build; confrontation and mobilization were beginning. The farm crisis had reached California.

The economic forces that had a direct bearing on the performance of the farm in the Central Valley were similar to those already confronted in Iowa. Although Central Valley farms were more diverse than farms in the corn belt, in reality much of the row-crop land could only profitably grow four crops. Cotton, especially, like feed grains in the Midwest, depended on the federal support program and on export markets. Thus by 1986, virtually all Central Valley row-crop farmers were at the mercy of the farm program and its uncertainties, and even the largest operators looked to support programs to raise their returns for cotton production to levels above those of the market. The same economic forces that made for greatest risk in Iowa replicated themselves in the Central Valley. Expansion to bring children into farming was a common response in the inflationary cycle, and those who chose this strategy, as in the corn belt, were more vulnerable than others. But in view of the perceptions of California agriculture from the outside, perhaps the greatest surprise was that large-scale operations were as vulnerable in the downturn as any. The conventional wisdom was that during a time of economic stress the bigger firm would hold up, but in fact, the giant agribusiness operations—the stereotypical bulwarks of California agriculture—often suffered more than smaller, family-sized competitors. The impact of the downturn was such that even the redoubtable Payne Company, which had been self-financed since it began operations in Corcoran in the twenties, was affected. In the fall of 1985 the firm offered 600 of its 1,340 salaried and wage employees a voluntary early retirement and severance pay package in an effort to slash operating costs. Even with this program, Payne took advantage of the farm program in 1986. Though its president complained that government policy forced the company to set aside 30 percent of its land, idling a large segment of its work force, like virtually every other large cotton producer in the state, Payne took maximum advantage of both government set-aside and cotton-processing programs.[6]

While there were similarities with Iowa, structural differences and the socioeconomic climate of labor-rancher antagonism prevented the kind of activism symbolized by the Iowa Farm Unity Coalition from burgeoning in the Central Valley. The activism of the sixties, exemplified by the United Farm Workers, had left a climate of suspicion it was impossible to dissipate. Even the churches were slow to mobilize. Farmers in the Midwest had a difficult enough time swallowing their pride and taking advantage of the welfare and mental health programs made available to them. In California, where welfare, mental health, and community action programs were geared mainly to minorities, their utilization by farm families in need was highly unlikely. Moreover, even in Iowa, where the Extension Service had a long tradition of "people support," the service was slow to appreciate the severity of the crisis. It is hardly surprising, then, that the logical institution to provide support for farm families in California, the Cooperative Extension Service, initially acted like a helpless giant. Ironically, early in the crisis it was the larger lending institutions, such as the Bank of America, that tried to help. Although the big lenders pursued highly publicized legal actions against borrowers, they also offered financial support for a hotline, and they lobbied in Sacramento for a number of legislative initiatives to assist the farm economy at the state level.[7] Finally, though it would be wrong to underestimate the seriousness of the farm crisis for agribusiness, it is important to reemphasize the diversity of the general economy compared to Iowa's. While many smaller communities were heavily dependent on agriculture and their support businesses—feed stores, farm supply, implement dealers—the dynamism of their retail trade shielded them against the kind of desperate soul searching typical of a similar-sized corn-belt community, whose isolation made it completely dependent on farm trade. Economic diversity was also of enormous importance to families thinking of leaving farming, for there were job possibilities in the immediate environment.[8]

For these reasons, therefore, the principal actors in the farm crisis in the Central Valley, while similar in many respects to those in Iowa, were different in one essential. There was no credible farm advocacy group without ties to an agribusiness organization concerned with the sociopolitical well-being of the farming community. In their time of travail, the California farm family was even more alone and dependent on its own devices, than was the case in the Midwest. Thus the California experience provides a test of the resilience of farmers and ranchers in an environment where farm failures were traditionally fairly common and where farm operators were encouraged to walk away and start all over again.

Farm Families

Farm families, characteristically silent and stoical, tended to hide any stress they felt, making it difficult to gauge the seriousness of the crisis in the Central Valley. Moreover, the kind of hard data routinely generated in Iowa—debt-to-asset ratios and other statistics—was totally lacking in California, for the Extension Service was oriented much differently. In Iowa the best state data on financial stress came through the annual farm business records survey, a venerable source going back to the early twenties. In California no such survey existed. The only reliable source remained the yearly USDA farm income and asset surveys, published many months after the data was gathered. This lack of data underlines once again the difference between the relatively "open" system of the Midwest and the "closed" environment of California, where it was thought business secrets would be disclosed if reams of statistics were released on the farm situation.[9] In a time of crisis, farmers and ranchers were even more reluctant to reveal their true position for fear that lenders would come to the wrong conclusions.[10] However, for all the lack of hard evidence generated by an independent body, newspapers and personal interviews provide enough material to draw conclusions about the chronology and major trends in California.

By the spring of 1985 some observers were suggesting that the economic position of California agriculture was no different from that of the Midwest. A commodities broker in Kings County painted a familiar scene when he claimed that farmers were all hurting; the only difference was in the extent of their pain. On a scale of ten, he rated at two or three those "who went out four or five years ago when land was worth $5,000 an acre, built a $500,000 home, bought land they could not afford, and got all their working capital screwed up. Those particular people are in trouble."[11] Stout denials of the problems, one analyst called the "redwood syndrome," noting that these trees seemed tall and powerful to the casual onloooker, but closer inspection revealed that they were hollow inside. Thus, California agriculture was impressive on the outside but hollow within. Indicators of crisis were all around. Bankruptcy filings were up in the first three months of 1985; banks and other lenders were beginning to accumulate considerable acreage, which, because of the poor state of the land market, they leased out rather than sell at depressed prices; and delinquency rates on agricultural loans from all types of lenders showed a dramatic increase. For example, by the end of 1984, 34 percent of Crocker Bank's non-real-estate agricultural portfolio was delinquent more than ninety days; the number of delinquencies for insurance companies rose from 22 to 47 between 1982 and 1984; 5.3 percent of Production Credit Association loans and 11 percent of Federal Land Bank loans were ninety days past due; finally the Farmer's Home Administration reported 53 percent of its borrowers delinquent on January 1, 1985, in the seven valley counties of Kern, Kings, Fresno, Tulare, Merced, Madera, and Mariposa.[12]

In this first brush with farm economic stress in California, the vulnerability of large corporations was evident. The huge Tejon Ranch, the Tex-Cal Land Management Company of Delano, which farmed vineyards and almond orchards in Kern County, as well as oil companies like Texaco and Superior, found themselves in severe trouble and sought to sell off and cut back.[13] Indeed, the favorite theme of lenders in 1985 was the resilience of the farm family in California compared to the giant corporation. According to the spokesperson of one major bank, farm families would survive "this thing," while corporations and businesses structured strictly for profit would go by the wayside, because "they could achieve greater profitability in other sectors." The well-managed family farm "had a lot more resilience and had a greater commitment to sticking with the business right through rough times." In short, the tradition of the frugal farm family was being summoned once again to pull farming through the hard times.[14]

The example of one Kings County larger-than-family farm helps explain why scale gave no advantage in a deflationary period. During the boom, this operator had expanded until he was farming twenty-six thousand acres. His troubles began in 1980, when the thirty-five-thousand-acre South Lake Farms came on the market with a $39 million price tag. In order to afford the property, the farmer went into partnership with two others. To raise his share of the purchase price, he refinanced his original property with the Land Bank, borrowing against the appreciating land values of the boom. "It looked like a solid deal made with three good farmers," said a Land Bank official. "Farm prices were good, real estate prices were increasing, and we had every reason to believe that farming would continue to be profitable." The partners secured additional loans for land improvement, equipment purchases, and crop production to the tune of $10.5 million. Unfortunately, the tide turned. By 1982 one of the partners could not make his mortgage payment. By borrowing money from the Bank of America, the first rancher made this payment but foreclosed on his partners. A year later the recession had driven down the value of his original property, further weakening his position. In October 1984 he filed Chapter 11 bankruptcy. His obligations included not only the Land Bank loan, but also $1.7 million from Bank of America, and $7.7 million from Producers' Cotton Oil. Because of the stake they had in the business, both Bank of America and Producers' were prepared to finance the firm for one more season, but the Land Bank, sensing a chance to recoup losses from foreclosure on the original property, tried to take back 22,500 acres. Fortunately for the operator, this request was denied by the court, and the farmer was permitted to continue to operate South Lake Farms, while his original property was leased out to local farmers by the Land Bank.[15]

Another huge Kings County operation, McCarthy Farms, which owed $40 million to Crocker Bank, also failed at this time, leaving a vast inventory of equipment. There followed one of the largest farm sales in county history.[16]

Large operations did have some advantages in liquidation, foreclosure, and work-out proceedings because more land and equipment were at stake and lenders were reluctant to absorb such large losses. Lenders were willing, therefore, to support them for a longer period. Large operators were also apt to have better professional advice on tax matters and legal questions involving the return of assets to the lender.

As in Iowa, farmers usually avoided decisive action at first, but the situation was more urgent by the spring of 1986. Smaller operators, especially the young or those who had brought sons into the business through ambitious expansion, often could not obtain the latitude and grace from the lender to work the debt out. For instance, the Coehlos, a

Kings County family that had expanded in the seventies, were foreclosed on by the PCA right after the grape and cotton crops were harvested in the fall of 1985. The family suffered through the winter without being able to resolve their differences with any of their lenders. As a result they had no operating money available with which to begin the spring work cycle; by April their grapes were overgrown, and the fields choked with weeds. This family took a do-nothing attitude, giving the initiative to the lender instead of exploring bankruptcy, rent back, or any of the other strategies successfully employed by farm families in trouble.[17]

A young grower in Fresno County, who had bought raisin land at the very highest price level of fifteen thousand dollars an acre in 1982 behaved more sensibly. After a weather-related crop failure and drastic plunges in price in 1983 and again in 1984, he realized his chances of getting financing for 1985 were slim. He went to the Land Bank on his own initiative because he could not make a land payment, and asked them what they wanted to do. The Land Bank foreclosed but then turned around and paid the grower a management fee to farm in 1985.[18]

Although the usual pattern was for the elder generation to succumb to the younger generation's desire to farm, with the result that both became enmeshed in severe financial problems, one California family resolved its intergenerational operation without this kind of trauma. When it became obvious that their elders refused to reevaluate rental agreements and reduce cash payments, the younger generation walked away from the operation to begin life again in the auto business. In this instance they had no debts, and they retained the house they lived in, for which they paid nominal rent to their parents.[19]

These cases, all of which occurred in the spring or fall of 1985, illustrate not only the breadth of the downturn but also some of the patterns typical of the farm crisis in its early stages, when farmers were not aware of their rights and lenders had not been stung by farmer intransigence. Although it might seem that in the business-oriented farm environment of California, farmers could be expected to take their failures in a "professional" manner, in fact they became just as litigious as Iowans had been. The surrender of a farm after years of work and several generations of ownership is an extremely traumatic and painful experience. Even several months after the PCA foreclosed on them, the Coehlos were still extremely upset, still mourning their loss; they were embittered by what seemed like the turncoat behavior of their business partner at the PCA and indignant that their friends and neighbors had shown little consideration for their circumstances. Here, of course, occurs one of the central dilemmas for families involved in economic stress: a decision to stand and fight, to take Chapter 11, might give the family a sense of purpose and a psychological lift; on the other hand, a workout or, better still, mediation with a lender might be more economically beneficial and save more of the business in the long run. By the funding season in the spring of 1986, some Central Valley families had entered a confrontational mode, and lenders could no longer expect passive acceptance of the situation.

But it would be wrong to suggest that litigation and confrontation were the only routes farmers and ranchers traveled in the early spring of 1986. Interviews with several families revealed that those with good contacts, a finely honed business sense, and determination to see their operation through hard times were rewarded with at least one more year in agriculture. Because of the uncertainties of the farm bill, signed by President Reagan in December 1985, they had to work harder to persuade their lenders to go along with them, and usually they had to return several times to work out funding details. One family gave up nine thousand rental acres to satisfy the stringent requirements of Bank of America.[20] Another farmer spent five winter months completely reorganizing the structure of his operation so that his sister, mother, and brother could take maximum advantage of the cotton program.[21] Many farmers had resorted to this kind of "creative activity" in an effort to stay in business. They formed dummy corporations with business partners to take maximum advantage of the deficiency payments. Each dummy corporation was designed to receive a fifty-thousand-dollar payment from the government in return for idling 25 percent of cotton ground. Whether such partnerships were legal according to the rules of the Agricultural Stabilization and Conservation Service was problematic. The important point is that such tactics measured the desperation among large farmers in the Central Valley in the spring of 1986.[22]

In the first quarter of 1986 confrontation had become more common between farmers and their lenders, and mobilization was beginning. "Horror" stories—accelerated foreclosures, the appropriation of savings accounts by lenders in lieu of loan payments, the raising of interest rates, and the signing over of land and equipment as a condition for financing—all were occurring and rapidly producing a "we-they" syndrome between farmers and lenders. In the Fresno area, raisin producers, who had suffered weather problems and price fluctuations, were particularly at risk. The profitability of the raisin business had attracted some families into full-time farming in the seventies. Part-time farmers had given up off-farm employment and devoted themselves to the farm.

Such was the pattern followed by the Shubins, who ran a small raisin and grape operation near Kerman. Shubin borrowed from the Land Bank to expand his ranch from 20 acres until in the middle seventies he owned 170 and rented another 30 acres. While this expansion had been done willingly, for Shubin was eager to leave his oil-field job, it was done on the understanding that both the Land Bank and Production Credit were interested in carrying only full-time farmers. Loan officers were not interested in "no growth farmers." Until 1983 they had "a super relationship" with their lenders, and the Shubins were told they could always borrow on their equity.[23]

When the bottom fell out of the raisin and wine grape business in 1984, Shubin could not come up with his forty-thousand-dollar Land Bank payment. He expected the bank to restructure his loan, but to his bewilderment, his loan officer called to say that they wanted to foreclose not only on his land but also on his house. Since he owed money to the PCA, and he needed their help for pruning, harvesting, spraying, and other operating expenses, he turned to them for assistance. They were prepared to carry him the following year, provided he signed over sixty-five acres and his equipment to them. He refused. Without financing, it was not possible to spray the vines, which succumbed to powdery mildew. Both his wine grapes and his raisins were severely damaged, cutting their value drastically. At one point, he pleaded with the PCA to lend him eighteen hundred dollars so he could spray for powdery mildew, but again they refused. Shubin decided to sue the PCA for damages, contending that the association had refused to lend the money necessary to save his crop in order to force him into foreclosure. Indeed, it appeared that the Farm Credit System was pressuring him into rolling over and leaving the farm. Fortunately the Shubins understood agricultural finance well enough to recognize the dangers of deficiency payments and obligations to the IRS, and so they decided to make a stand. Like thousands of other farmers, however, they experienced a sense of guilt, feeling that somehow it was all their fault, that the nightmare might never have happened had they done things differently. Shubin, like others before him, partially exorcized these feelings by walking into his lender's office and telling the loan officer to try something "sexually feasible only among certain fish and lower invertebrates." It made him feel somewhat better.[24]

The Costas, who farmed near Los Banos in Merced County also found their lenders to be fair-weather friends. Costa had worked for years in a management position on one of the corporate farms on the west side. By 1979, he had accumulated enough capital to start out on his own, and his lender, the Bank of America, was good to the family, sponsoring rapid expansion until Costa was running a fifteen-hundred-acre operation, of which two-thirds was leased. With land bought at twenty-five hundred dollars an acre, much of it in scattered locations, the Costas were vulnerable to a downturn. Trying to buy high-priced land, which was losing value each year, and renting property with leases geared to inflationary times rapidly destroyed their dream of succeeding as independent operators. They lost their leases and equipment and only managed to hold onto the five hundred acres they owned by filing a land patent. This strategy, common among desperate farm families, had little value other than to cloud the title, but it was a measure of the Costas' desperation and bitterness. The Costas, "pleasant, articulate, and industrious people," had spent years scrimping to finance a dream that evolved into a nightmare. Their lawsuits were the result not of trying to walk away from debts but of trying to get back into farming, to be given another chance.[25]

Without a farm activist organization readily available, it was difficult for families like the Costas and the Shubins to break the barriers the highly individualistic California farmers put around themselves when they were under stress. In Fresno County, newspaper reports of American Agriculture Movement activity did surface, but apparently this organization was unable to gain grass-roots support as a voice of ranchers. Without a doubt, the diversity in type and size of farms kept farmers apart and had a retarding effect on activism. Larger farmers had such strong connections to the lending establishment, moreover, and the corporate attitude was so thoroughly inculcated among them that they were unlikely to bite the hand that had fed them.

The Lindemann family, for example, whose German immigrant grandfather had arrived in Los Banos from Wisconsin in 1928, gradually built up a sizable farming operation, which included a dairy and row crops. When the middle generation was killed in a plane crash in 1968, two sons expanded enormously, leveraging their way to a thirteen-thousand-acre empire on the west side, with equipment worth $8 million. But by 1983 they too were incapable of servicing their debt and so began a three-year program of liquidation. Big farmers, it would seem, could engineer a workout with their lenders. In this instance, Wells Fargo took most of the Lindemanns' property in a complex deal that left one brother with a packing plant and the leasehold on the former home ranch. The other brother left farming altogether and took up a position as an agricultural consultant. Among his duties was helping other farmers to restructure their debts. This kind of pragmatism was also shown by a Corcoran family who had farmed in the area for over eighty years. By the end of 1986 they decided to liquidate their operation before indebtedness overcame them. They were motivated partly by the serious illness of a family member and partly by the specter of a neighbor who had lost everything in a vain attempt to save his farm. One brother sought a position in an agribusiness firm; the other decided to sell real estate and farm on rented land elsewhere.[26]

The Lenders

One might assume that the giant multibillion-dollar agricultural lenders of California, with their reputation as progressive corporate institutions, would not face the agricultural farm crisis in their state in quite the same way as an independent institution with assets of $40 million in the Iowa corn belt. However, when farmers were faced with going out of business and lenders were prodded by the bank examiners and their corporate leadership to strip their portfolios of nonperforming loans, neither side was inclined to behave in an enlightened fashion. For farmers as well as bankers, it was easier to apply a strategy for expansion in an inflationary climate than to plan workouts, foreclosures, and liquidations in a period of deep recession.[27] It was obvious that the large California banks had done some clumsy planning and made some unfortunate decisions. One bank analyst described Bank of America as "a mismanaged, very troubled organization," run by a staff who did not think ahead. With its hundreds of branches, it was a bureaucracy "out of control." Not surprisingly, in the confrontation phase, which was punctuated by a series of highly publicized lawsuits, this giant corporate organization was unable to avoid the wrath of borrowers who were determined on litigation.[28] By the spring of 1985, when the bank's own figures indicated losses of $82 million from agriculture in the previous year and continuing difficulties in the first months of 1985, the agricultural corporate staff of Bank of America moved into a survival mode.

Vice-presidents took up the cry of "thrift," which was being trumpeted across the country. As one spokesperson suggested, profitability could return to agriculture only through sound basic business practices. "You cannot rely on the business principles of the seventies to manage today," he declared. "We can no longer be content to have a carryover or loss this year because things will be better next. You just cannot manage that way." In reaction to the economic distress caused by deflation, a strong dollar, and the loss of markets abroad, agriculture was going through a "true structural change." Farmers would have to adjust to the lower prices on world markets, and profit margins would be greatly reduced. In order to get by, the good manager needed to assess the what-if probabilities in his business and develop a minimum of three alternatives to cover every eventuality. Under the old way of doing things, risk was usually managed through land equity. In the new business climate, it was necessary to understand the realities of the marketplace to compete. Farmers in sound shape "understand markets; they understand sound basic financial business management principles where working capital is needed, where debt-to-asset ratios are controlled, and where if a crop gets below break-even costs, you can't plant."[29]

But while corporate philosophy sounded logical and reasonable, in the branch banks throughout the San Joaquin Valley a panic policy was instituted. Liquidators with a talent for ax wielding were brought in. Much of the local decision making was taken out of the loan officers' hands; instead, computers, manuals, and checklists were utilized to make decisions. To quote from a story published in a San Francisco newspaper, which could have run in the Des Moines *Register* without a single word change, "City folks in pin-stripe suits replaced the local banker who knew farming and his customers. The loan officer who worked with families and their farms for years was isolated to a special task force on problem loans, replaced by a newcomer with a computer."[30] Bank of America had a twenty-point list for evaluating farm loans. If one check mark went into the wrong column, the borrower had to look elsewhere. Loans would get bumped up to higher authority and often into limbo, and the farmer would not hear a word for months on end.[31]

As in the Midwest, the confrontation between lenders and borrowers was probably inevitable. Both sides needed to work through this phase before the economic and psychological costs could be seen as too great and mediation could be put in perspective as a useful tool.

By the summer of 1985 the Bank of America had decided on new tougher policies for the coming growing season. An agricultural loan preparation guide published in September of 1985 was issued to customers to "assist in developing efficient management information tools." The preface pointed out that the concepts in the guide were nothing terribly new. There was a quotation from the 1918 Bank of Italy "Farmers' Account Book," which expressed the bank's age-old policy of trying to make California farmers more "businesslike": "The successful farmer is one who studies his business. . . . The time has come when the farmer must manage his farm upon a systematic basis, just as the businessman his factory or store. Growing competition and progressiveness demands that the farmer shall know which products of the farm are bringing profitable returns and which are not." There followed a fairly standard application form, listing the documentation requirements for a Bank of America loan, followed by a detailed budget form for income and expenditures over the next eighteen months. Like other lending institutions, the bank required, among other things, tax returns, financial statements, leases, crop inventories, insurance policies, estate plans, plats, details of landownership, and a listing of equipment, with serial numbers—about the same information required by, for instance, the Visalia PCA in 1947.[32]

Obviously the charge that lenders resorted to more stringent requirements to weed out poor managers would seem rather hollow in view of the traditionally businesslike accounting methods in the Central Valley. Many argued, however, that the strict guidelines would eliminate the innovator and the experienced grower who knew where his costs were but could not put them down on paper. Even the head of the august California Farm Bureau Federation seemed to endorse this view when he suggested that lenders were being overly conservative: "I think what will happen is some people who are good managers and probably have sufficient assets . . . will be questioned or denied credit because these people don't fit into that computer mold."[33] In the final analysis, it was not the imposition of computers or a blizzard of paper work that cast a shadow on the relationship between farmer and lender but rather the lenders' perception that true structural change was occurring in agriculture and that lenders and borrowers had to hurry to catch up. In such a revolution old ways of doing business had to be jettisoned, and the trust implied in the practice of sealing a deal with a handshake was no longer applicable.

The sea change in California agriculture between the "glory days" of the late seventies and early eighties and the depression of the middle eighties is classically illustrated by two families whose lawsuits against the Bank of America received extensive publicity. Both families farmed small to medium-sized operations near Woodland and Yuba City, an area of California that still had some affinity to the Midwest in the way agriculture was organized and in the attitudes of farm families. The Furlans, who had migrated from France in the 1950s had always depended on the Woodland branch of the Bank of America for financing but until 1980 had kept their operation small. The Stanghellinis, a well-established third-generation Italian family, had never required operating money until 1980, when the Bank of America persuaded them to take out a crop loan to finance expansion on rented land.

The evidence suggests that both families were better growers than business people and would be prime targets in any situation where a powerful lending organization wanted to recoup some of its losses at the expense of innocents. For example, the Stanghellinis kept $800,000 in a passbook account paying 5 percent interest, while interest rates on other kinds of investments were in double figures; meanwhile they financed their ranch with high-interest loans. Both families put their trust in the bank—which preached leveraging, expansion, and the utility of its computerized agribusiness planning service—instead of practicing old-fashioned caution, which had served them well in the past. Both farms had been conservatively managed businesses with little debt before the eighties. By the end of 1984 they both owed the bank over $3 million. They had suffered tomato crop failures and poor prices, and the bank decided to call in their loans.

In the debacle that followed, the bank acted brutally, using questionable practices to recoup losses on other loans. The lender forced the Stanghellinis to put up their almost-paid-off home as collateral and then doubled the interest rate; later the family had to hand over a savings account as a condition for receiving a loan, only to have the bank cancel the loan. The Furlans sold land to pay off some principal and interest on a loan to improve their debt-to-asset ratio, but the bank nevertheless refused to work with them. Foreclosures, Chapter 11 filings, and civil suits brought both families to the courtroom, where they faced the bank's lawyers. Their cases rested on the charge that the bank had defrauded them by acting like a friend while pursuing a course to ruin them and take their land—"by giving with one hand and taking with the other."[34]

As in hundreds of other cases throughout the nation, there was an imbalance in leverage between the borrower and the lender. Once farmers signed up for loans, they had to follow the guidelines spelled out by loan officer and computer. Thus, the bank effectively took over as manager of the farm. From the bank's viewpoint, it was a case of investing in a business proposition and then not getting paid; the borrower had signed a contract and not abided by it. However, such a view could do neither party much good. It was a phase in the farm crisis that had to be weathered.

As the institution with the largest agricultural portfolio in California, not only was the Bank of America at greater risk than other lenders from frustrated customers, but it also had more interest in moving on towards more productive solutions. "The fact remains," said a spokesperson, "the Bank of America not only sympathizes with the plight of California farmers, we share their plight."[35] In view of the huge losses of the institution, such words were especially to the point.

As in the Midwest, the Farm Credit System remained in the background in the initial stages of the crisis, but as the crisis deepened, the Land Bank and especially the PCA were active in foreclosures. In general, however, the California system was in better financial shape and far better managed than that of the Omaha district. For this reason, the California Land Banks kept the loyalty of their members. The national reorganization in the final months of 1985 had an impact on the Sacramento District, for it ensured that funds were transferred to other areas of the country, where conditions were not so bright. The membership did not appreciate this turn of events, but there was little that borrowers with debt problems could do about the situation.[36]

For a variety of reasons, then, lenders, especially commercial lenders, were in the forefront of activity in the early phases of the farm crisis in California. The size and sophistication of lenders did not seem to change the way the crisis developed; the same mistakes were made by the giant California institutions as were made by tiny Iowa banks. One other conclusion seems merited. It would seem that although the largest farms were as vulnerable to the downturn, lenders were more willing to negotiate with them for a workout than with smaller operations. For example, rumors about the downfall of the Wood operation in Corcoran were rife in the winter of 1985-1986. The purchase of millions of dollars of lake bottom land as recently as 1979, the family feud, labor troubles; a transition in management, and flooding had weakened the firm even before the downturn. Oil revenues from wells at the west end of the Tulare Lake bottom, which had provided the corporation with much-needed off-farm income, were drastically reduced. By the early spring of 1986, however, the Woods had secured funding for another year. The Bank of America had given the firm operating loans that totaled an astonishing $82 million. Apparently, this backing was obtained partly on the strength of the corporation's desire to move into vegetable production in other areas of California. Unfortunately the Bank of America was so hard pressed by the summer of 1987, that the Woods were forced to drastically reorganize their vast operation. As many as 40,000 acres were given back to the bank, after it refused to provide operating money for the following year.[37]

The Establishment

In November 1985, the USDA Extension Service published a survey of state Extension Service involvement in programs to assist distressed farm families. Compared to midwestern states such as Missouri, North Dakota, and Iowa, California Extension Service involvement was minimal, partly because there was less need, partly because many still believed the state's unique mix of commodities and the structure of its agriculture would protect it, and partly because of the way the Extension Service itself was funded and organized.[38] The California Extension Service, unlike other state organizations, is decentralized. Regional and even county components are autonomous from the administrative centers at Berkeley, Davis, and Riverside. The agency also has very close ties to agribusiness, which, especially in the sixties and early seventies, caused much controversy.[39] Whereas in 1983 in a state like Iowa only 40 percent of funding came from state appropriations, California provided more than two-thirds of the service's $100 million budget, with $30-35 million going for research alone.[40] Not surprisingly, this money was spent on projects that benefited the service's most visible constituency, the larger growers. In addition, Cooperative Extension also received impressive support from private and corporate sponsors. Over the twelve-month period from July 1, 1978, to June 30,

1979, a total of $3,121,396 was donated by such sponsors as Dow Chemical, DuPont, Chevron, Hunt and Wesson Foods, the Gallo Education Foundation, Adolph Coors, Desert Cottonseed, Christian Brothers, and the Walnut Marketing Board—virtually a roll call of California agribusiness—as well as many individuals who gave smaller amounts.[41] When the Giannini Foundation of Agricultural Economics at Berkeley, founded many years ago with support from the Bank of America, is also included among the corporate sponsors of agricultural research, it seems clear that the mission of the California Extension Service would have a somewhat different emphasis from that of its less well endowed counterparts elsewhere.

In Iowa, once it mobilized, the Extension Service had trouble attracting clients to its programs not because it was doing anything fundamentally wrong but because potential clients continued to feel uncomfortable with the idea of seeking help. They feared that they would be exposing their own inadequacies, and so, for all their discomfort, they preferred to deal with their economic and psychological problems themselves. In California it was the decentralized system that kept the Extension Service ineffective. Though it could not play an advocacy role, the agency might at least have measured the extent of the crisis, but there was a leadership void caused by decentralization. By the spring of 1986, when the legislature had before it a dozen or so initiatives aimed at alleviating the situation, there was still no data with which to gauge economic distress except that generated by lenders.

These circumstances underline the importance in a democratic society of an independent watchdog on the functioning of that society. Because there was no such organization in California, there was little incentive for the Extension Service to explore nontraditional avenues of service or to assume leadership in unfamiliar areas. Interviews revealed that some farm advisers were acutely aware of the problems some of their clients were having with economic and psychological stress. They made efforts to run financial management seminars that dealt with such topics as the implications of Chapter 11 and the nuts and bolts of management in a period of austerity. However, poor attendance discouraged the organizers. Though extension agents perceived the need for one-on-one consultation—frequently desperate clients broke down on the phone while talking about their problems—the pressures of their regular work, the lack of support from the Extension Service itself, and the extremely sensitive nature of the issues involved in counseling prevented effective efforts. In the summer of 1986 the Fresno and Tulare extension offices did survey all ranchers in their counties in an effort to gauge the severity of the downturn and the needs of farm families suffering economic difficulties. A poor return rate damaged the statistical validity of the survey, but it did reveal that 48 percent of all respondents were "struggling" and that less well established medium-sized farms were in greater difficulty than others.[42]

During the spring of 1986 a review of other programs put on by the California Extension Service could be described as business as usual. A management seminar in Tulare County in March, jointly sponsored by such organizations as the Farm Bureau, the major banks, the Farm Credit System, and California Women for Agriculture, carefully skirted the troubling issues of the time. Instead, the audience was given a series of upbeat presentations that stressed the corporate view that austerity and efficiency would meet the challenges of the day.

Church involvement was essential in producing farm support systems in Iowa, but churches in California failed to respond. According to one observer, the churches were doing little in practical terms to help farmers in crisis. A University of California Extension Service social worker claimed that the farmers she had talked to had expressed a sense of abandonment: "The feedback that I have gotten is that they're not getting anything from anybody." The slowness of the churches to mobilize had fairly obvious roots. With rare exceptions, such as the Mennonite congregations in Tulare County, churches were almost entirely urban institutions, with little or no concentration of rancher membership. Although congregations were sympathetic to the problems of distressed farmers, few saw the need to go out of their way to help. Moreover, the rancher-labor dichotomy made some churches ambivalent. The division was so sharp that many doubted whether laborers and ranchers could be accommodated in the same churches, and they wondered whether their ministry was among ranchers under stress or among exploited laborers.

In the San Joaquin Valley, several Catholic dioceses had clearly chosen the latter. The Fresno diocese, for example, set about obtaining scholarships for minority youth in the state's agricultural colleges. Another program sought a dialogue with growers about "responsible capitalism"—better working conditions for laborers and a reduction in influence of the *coyote*, the notorious contractor who abuses his workers. In Stockton the diocese attempted to assist small farmers with its own small farm viability project. It operated a demonstration farm and tried to encourage minority farmers to improve their skills in small-scale vegetable production.[43]

Social service agencies, too, were organized more for a working-class urban clientele than for a rural business class. The many excellent not-for-profit organizations that dealt with large numbers of needy people were willing to devote some of their time to the wants of farmers, but they had no expertise in rural assistance. To take one example, a Fresno organization, Help in Emotional Trouble, which ran a counseling and referral hotline to help rape victims and suicides, was ideally placed to serve the emotional needs of farm families as well. For several months in the winter of 1986, this agency waited for the Extension Service to organize training for financial counselors with money donated by the Bank of America. It was important to get such a service in place before farmers began seriously looking for operating money for 1986. The hotline never materialized in 1986. Only in the winter of 1987 did the Extension Service in the southern San Joaquin finally initiate programs that were commonplace elsewhere.[44]

At the grass-roots level, then, California farmers under emotional and financial pressure were left to their own devices for a long period. Without prodding from outsiders, farmers remained reluctant to expose their tribulations in the early stages of the crisis. For this reason the period of denial and confrontation was a lonely experience for many, who found support only from a few close friends and relatives.

While the California crisis moved to its own timetable, attorneys were able to learn from mistakes made elsewhere, particularly the Chapter 11 scenario, which was used so liberally and with such poor results in Iowa. California farmers had a healthy skepticism of Chapter 11, mainly because of their reluctance to invest in the kind of expense involved. At the same time, lawyers themselves were better acquainted with the ramifications of Chapter 11 and the chances that a farm could usefully use it as a tool to reorganize.[45]

The press emphasized the steep increase in bankruptcies and pointed to the faltering farm economy as the reason, but in fact, the difficulty of identifying farm filings in the bankruptcy court made it hard to confirm whether farmers were filing in great numbers. There were ninety "farm-related" bankruptcies in the whole of the southern San Joaquin in 1985 and twenty-seven in the first three months of 1986, but it is unclear whether this statistic included agribusiness firms. At any rate, compared to Iowa, these numbers were small and seemed to confirm the more sophisticated attitudes towards bankruptcy among attorneys and farmers alike.[46]

Actually, bankruptcy was becoming more common among the general population. The press attributed the increase in nonfarm bankruptcies to lawyer advertising and the low cost. They also stressed there was some abuse of the system, especially Chapter 7, by those who were not interested in working out their debt problems.[47]

The evidence suggests, then, that most California farmers were accustomed to conducting routine business transactions which required legal representation. Their more formal way of doing business made them in theory better able to deal with lending establishments and militated for resolution through negotiation rather than litigation.

The size of firm had a marked bearing on the outcome. Large firms that retained lawyers on a permanent basis would be strongly inclined to resolve differences in a rational and businesslike fashion; smaller firms did not have such props. In practice, however, this was not necessarily so. In the spring of 1987 a long-drawn-out case involving a large Castroville farmer who in 1983 sued Wells Fargo bank for fraud, bad faith, and breach of contract after the lender withdrew credit, was settled in a jury trial with an award of $60 million for the rancher. The case created a precedent for lender liability in California agriculture, and although the bank appealed, the victory for the farmer had considerable significance. For not only would it encourage others to pursue litigation, it would inevitably cause lenders, who were already leery of ranching, to be even more cautious with agricultural loans in the future.[48]

The Ecological Crisis

In California the farm crisis involved a threat that was even more disturbing than structural change in the agricultural economy. On the west side of the San Joaquin there was a danger that farming would have to cease altogether. Both state and federal authorities were intimately involved in a situation that came to a head quite suddenly in the spring of 1985, a situation that had been festering for twenty years while large sums of money were spent to transform the desert on the west side of the valley into some of the most productive farmland in the world.

The idea that the Central Valley would someday go the way of many hydraulic societies of the past and that nature itself would cause a major reevaluation of irrigated agriculture was not seriously contemplated before 1985. For years, the major priority for ranchers was the delivery of water from the Sierra to take advantage of ideal growing conditions with relatively cheap water supplied by massive irrigation projects financed by Washington and to a lesser extent by Sacramento. As more and more desert soil came into production, warnings of perched water tables and salt buildup were ignored in the face of bountiful crops.

The problems of salts in the soil and an impermeable clay layer that trapped water and created a perched water table were known from the time the first irrigation water was pumped on the west side. It was also known that as more ground was irrigated, the perched water table would continue to rise, pushing the salts toward the surface. The solution to the dangers of salt buildup was the construction of tile drainage, which would move the used irrigation water elsewhere. In the northern San Joaquin many thousands of acres were tiled in this way, but in the Westlands Water District only about eight thousand acres were tiled. More to the point, the major drain, which was to have transported the bulk of the waste water northwards to the San Francisco Bay, was never constructed because of opposition from environmentalists and local governments. Instead, the drain was completed only a short distance northwards and deposited its wastes into what was intended as a regulating reservoir. In effect, the wildlife refuge at Kesterson became a huge evaporation pond. It was there, in 1983, two years after the first drainage water began flowing, that dead and deformed bird embryos were discovered.[49]

The birds were found to contain high levels of selenium, one of the naturally occurring elements leached from the soil. After considerable behind-the-scenes maneuvering, a halt was ordered in the delivery of federal irrigation water to forty-two thousand acres of land in the Westlands District in March 1985. In the immediate area of the closure, the reaction was both angry and incredulous. "Sure we'll be killing a few ducks," remarked a farm equipment dealer, "but we will be killing a lot more than that if this shutdown sticks." Most farmers had already borrowed considerable sums to prepare for a new year of cantaloupes, tomatoes, cotton, and other crops, and a ban on farming with so little warning would have been an unmitigated disaster. Fortunately, important politicians intervened and persuaded both the Department of the Interior and state authorities to stay the order so that all parties could work together to find some permanent solution to the ecological problem.[50]

One of the problems with this situation was that no one was prepared to pay the economic and environmental price for the privilege of employing large-scale irrigation agriculture. After Kesterson, the outlook for west-side farmers already in trouble because of deflation, overproduction, increased water costs and low prices, looked especially grim, for long-term solutions to the problems of salt and toxic waste would be expensive, both to the individual farmer and to federal and state government. In all probability this expense would make agricultural production only marginally feasible.

One possible solution for the individual farmer was the construction of on-farm evaporation ponds. For every twenty acres drained, one acre would have to be set aside as an evaporation area. Obviously, construction would be expensive and, in view of the reduction in productive capacity of a given farm, hardly worth the trouble. All the other solutions involved massive construction projects and the expenditure of millions if not billions of dollars. At a time of extreme austerity in the federal budget, it was unlikely that a proposed $14 billion pipeline to drain irrigation water into Monterey Bay would be politically, economically, or environmentally feasible. In effect farmers in the area of the shutdown were left in limbo, while the water engineers, lawyers, politicians, and bureaucrats attempted to find a solution to a problem that really affected not just the limited area that drained into Kesterson but ultimately 600,000 acres of the west side.[51]

A year later neither the federal government nor the state of California had found a solution. The drains from the forty-two thousand acres were plugged to prevent spent irrigation water from flowing into Kesterson. Moreover, by the spring of 1986, the situation had begun to affect the nonfarm community as well. Two valley communities, as well as Fresno County itself, had taken legal action to prevent the spread of selenium-tainted groundwater into wells and sewage systems because of the plugged drains. Although all parties eventually agreed to cooperate with the federal and state authorities, pending the outcome of tests for selenium in the groundwater, it appeared that the selenium issue—which for the moment affected only a relatively small area—might evolve into the kind of nightmare against which any amount of bankruptcy and economic stress would pale in comparison.[52]

It could be argued that the west side of the San Joaquin Valley should never have been brought into production in the first place. The millions of dollars spent on irrigation projects benefited a few hundred farmers and giant corporations, and most of the food produced was surplus that lowered market prices. On the other hand, there was some merit in the argument that after spending millions on the projects, government could not just walk away and leave the huge acreage to revert to salt pan. To do so would be to deal a crippling blow to the agricultural economy of the valley, if not the state as a whole.

Ironically, the threat to the livelihood of farmers directly affected by the plugging of drains in Westlands stirred them into cooperative action to try to save themselves. Growers, farmworkers, businessmen, and city officials formed themselves into the Community Alliance for Responsible Water Policy, with 273 members. They drew up a three-year water plan designed to drastically reduce irrigation. It was a stopgap, which, together with a change in the start of the district water year from January 1 to March 1, would allow growers to forgo preirrigation and utilize rainfall to a greater extent than before. The plan was hurriedly put in place to save farming for one more year. According to their legal counsel, the growers, with their backs to the wall, had no choice but to solve their own problems, because the government, whose responsibility it was, had failed to find a solution.[53]

Conclusion

California's reaction to the farm crisis was enigmatic. *California Farmer*, for example, the leading farm publication, often gave greater coverage to the downturn than did the midwestern trade papers. On one occasion it even put a large color photograph of a farm family fighting the Bank of America on its cover. Yet at the same time, the Extension Service preferred to keep a low profile. Cotton farmers, who only a few years before had scorned government support in agriculture, rushed to take every possible advantage of the program in 1986. The dairy buyout scheme, which was designed to encourage small midwestern dairy farmers to leave the business, was utilized as much by Californians as by farmers in the upper Midwest. It seems that free-market ideology was ignored when growers faced delinquency.

For California, with its huge, diverse economy, the farm crisis was a localized "deep recession" that made little impact on the average citizen. Moreover, the organization of agriculture itself was so oriented towards the corporate business ethic that, despite the symptoms of economic stress, there was little mobilization to deal with it either by farmers or by advocates on their behalf.

Nothing illustrates the difference between the Midwest and California better than attitudes towards the electronic media. At one farm meeting in Fresno in the summer of 1986, a rancher activist had arranged for television coverage. Before the meeting he was being interviewed by a reporter outside the hall. He noticed a number of ranchers drive up and then drive away when they saw the cameras. Later, he received calls from growers expressing indignation that he had allowed the media access to the meeting. Apparently, they feared that if their lenders saw them on the television news, they would consider the ranchers economically vulnerable and would cut off operating funds.[54]

In California farmers and ranchers, it would seem, were expected to act like businessmen and to accept failure as routine—to walk away and go through a career change like any other business executive. Another fundamental difference from the Midwest was in the area of political action. In Iowa some of the most effective political action on behalf of farmers was carried out by independent citizens' groups; in California, however, lobbying for state tax relief came from the banks and the establishment farm organizations. It is significant that in California, where there was a budget surplus, tax relief, not a moratorium, was deemed a priority. Beginning in May 1986 farmers received state income tax reductions slated to continue for the next several years.

In the summer of 1986 the farm recession was far from resolved. Interviews with family farmers reveal that beneath a brave exterior many ranchers were suffering great financial stress. In addition, the ecological problems of the west side, which threatened a comparatively large number of farm families, laborers, agribusinessmen, and communities, posed perhaps an even greater challenge to the future of large-scale California production agriculture.

Conclusion

Whither the family farm? Is it possible to discern a major watershed in the structure of American agriculture in the middle eighties using the evidence assembled in two important farm states? The short answer would have to be "not yet." The historical precedent of the 1920s and 1930s seems to suggest that change is unlikely to be rapid. If the experience of those decades is largely irrelevant from an economic and structural point of view, something can be learned from what might be called the mechanics of recovery in those days. The psychological impact of the Depression on farmers was deep, and recovery was slow. Likewise, present-day chances for a rapid transformation of agriculture seem remote.

Fortunately those concerned with the preservation of the family farm and the promotion of a more sustainable form of agriculture can work with a blueprint provided by government researchers and plan accordingly. A possible scenario for the next twenty-five years is mapped out by the Office of Technological Assessment's *Technology, Public Policy,and the Changing Structure of Agriculture.* At first sight, the prospects for the medium-sized farm with gross sales of between $100,000 and $250,000, where the operator works full-time on the farm, do not appear bright. In the brave new world of high-tech agriculture, where the wonders of gene splicing and embryo transfer allow dairy farmers to produce more milk with fewer cows and where the marvels of microbial inocula and plant propagation permit the farmer to grow disease-free plants, the advantages of the super farm would seem to be overwhelming. According to the study, the large-scale firm would be highly efficient in production, marketing, and financial and business management. Such operations would be run by "full-time, highly educated business managers. Barring unforeseen acts of nature, farm operators would be able to predict their chances of making a profit before planting or breeding."[1] The medium-sized farm, the backbone of American agriculture, would be less able to compete, partly because it lacked access to the information and finances necessary to adopt the new technology effectively. Indeed, all the research in the early eighties on which the study relied showed that large farms had advantages over smaller ones. Emergency credit programs, soil conservation, and research at land grant colleges—all tended to benefit the large farmer. Similarly, government tax policy and the ubiquitous farm programs had a major impact on increasing farm size by assisting large farmers at the expense of small. For these reasons, according to this report, without "substantial changes in the nature and objectives of farm policy," medium-sized farms would gradually be eliminated. They would be replaced by just two types of operation: the small, part-time hobby farm and the large unit, often run by a manager, with sales of over $250,000 in 1982 dollars.[2]

Obviously the consequences of such change for rural America, especially in the Midwest, would be considerable. It would, for example, guarantee further deterioration in the infrastructure of small communities. At the same time, a vision of high-tech agriculture dominated by large, often corporate-run farms, did have a certain vulnerability. The biotechnological revolution in farming was controversial because it would only compound a major problem that faced agriculture throughout the twentieth century—chronic overproduction. The idea that the United States could sustain its agricultural overproduction through exports by adapting high technology would seem to be an illusion. High technology knows no natural boundaries and would be quickly adopted by others. Likewise, the take-over of agriculture by large-scale firms would seem less inevitable if they were forced to complete with smaller operations on equal terms. If government programs were eliminated for firms grossing over $250,000 per year and they were also denied access to low-interest government loans, the medium-sized family farm would find itself in a more favorable position.

Thus, there is a possibility that the medium-sized family farm could rise from the ashes of the farm crisis. One hopeful sign is the climate for change in the position of the farm family vis-à-vis its business partners. For years, the farm family was held captive by outside forces. Like so much putty, it conformed its management, operation, and ideology to forces that, as the debacle of the past decade showed, did not necessarily have its best interests at heart. If the farm crises has any value, it is in the area of consciousness raising. Farmers now have a chance to take authority for their own lives, to make their own autonomous decisions. If they have learned this lesson, never again will finance, law, politics, and community pressures find them so acquiescent and powerless.

The ability of the medium-sized family farm to survive grows out of its unique organization, which augments the tendency of farm families to cling to the land. Although there has been a trend toward the differentiation, concentration, and transformation of farming, and this process has been accelerated by economies of scale and government policies, there are other factors inherent in the structure of agriculture that favor the family farm. Risk, internal diseconomies of scale, the immobility of fixed capital, and the self-exploitation factor—all combined against the transformation of most of agriculture into the kind of farm organization found on the Tulare Lake bottom.[3] In other words, although the farm crisis put greater emphasis on farm management in place of tractor driving, much of farming remains hard and dirty work, best undertaken by families who can undergo the risks and are willing to accept the relatively low returns for their labor. In a corporate-dominated world, well-educated managers would be unlikely to accept the poor working conditions and low returns typically found in agriculture.

If the medium-sized family farm maintains an important position in farming in the future, its place in the tenure system will be affected by the shake-out of the eighties. Patterns of recovery might well follow those of earlier decades. For instance, in the twenties as in the eighties, it was the so-called progressive farmers, who expanded in land, buildings, and machinery, who took the hardest hits. Some evidence indicates that their places will be taken in the Iowa corn belt not by absentee corporations or syndicates of lawyers and doctors but by resident farm operators who were conservative in the boom times and waited to buy land at the lowest possible price. Grass-roots land tenure is like a game of musical chairs: if one operator stumbles, another, usually a neighbor, is there to take his place. In the resolution phase of the farm crisis, mediation and the process of allowing the system to work itself through enabled some farm families to lease land back from the creditor. As in the thirties, liberal credit polices could permit those who lost farms to reestablish themselves. Given the unavailability of nonfarm employment in rural areas and the deep impulse of many rural people to operate their own farms, there will always be a pool of potential operators to take advantage of whatever opportunties are presented.

Another small but important group of prospective purchasers are those natives of a particular area who moved away for higher education and entered other occupations. These professionals and business people are concerned about their home communities, are knowledgeable about farming and rural life, and feel they can contribute to the restructuring process by purchasing farmland and renting it out to resident local operators. This pattern also occurred in the late thirties and forties.

Legal restrictions will prevent banks, insurance companies, and the federal credit system from controlling vast amounts of foreclosed farmland and placing it in the hands of farm management firms, at least in the Iowa corn belt. There are forces, often in the community itself, however, that would welcome a slackening of the laws to allow agribusiness a more permanent foothold in landownership. Before the farm crisis, land tenure was a battleground between the neopopulist forces who wanted to preserve the family farm and the advocates of corporate financial intrusion, who desired opportunities for off-farm investors. There is every reason to believe that this battle will continue and perhaps intensify over the next several years. In California the future of land tenure arrangements is obviously much more complex. Investor ownership has been an established practice for years. However, the evidence suggests that big farming has never been more vulnerable, and opportunities for a more broad-based tenure system could develop because of the uncertainty of agriculture as an investment for off-farm interests. In sum, a shake-out in land tenure means that there will be a rearrangement in landownership. It will be important for the medium-sized full-time farm family to remain mobilized to protect its interests.

The prospects for the multigenerational family farm are tied to this working-out process. Possibly the most magnificent irony of all in the period of boom and bust was the effect of the Economic Recovery Act of 1981 on the family farmer. Part of this legislation was specifically designed to make it easier for the farm to be passed from one generation to another through cuts in inheritance tax rates. Yet though the intention was to ease the burdens of transfer, in four years the massive income tax cuts included in the act had contributed to the slump in the agricultural economy by cutting inflation. Farm families no longer fretted about estate plans; they worried about strategies to save the farm itself. Bankruptcy was not a legacy to pass onto the next generation.[4] Multigenerational farming partnerships received a strong setback. Moreover, the downturn maintained the relative advantages of the senior generation in ownership and management, for the tax laws discouraged intervivos transfers of land to the younger generation. In addition, in the wreckage of the Farm Credit System, younger farmers usually needed the signature of an elder family member on a line of credit. The multigenerational operation, then, remains, as it always has been, a difficult form of business organization to pursue.

One argument strongly against the take-over of agriculture by larger entities, at least for the near future, is the comparative lack of funding available for agriculture, either because of heavy losses or because of a reluctance to risk loans. Indeed, the health of the whole agricultural credit system—especially the federal system—remains troubling, and its recovery uncertain.

Ironically, leaner, more efficient operations could translate into a more self-sustaining agriculture, less dependent on expensive inputs. In this event, the downturn could have benefits for the environment. Some progress has been recorded by the government acreage set-aside program. Recent Federal tax revisions could also be of great benefit in eliminating tax-loss farming. In turn they would boost Iowa livestock production and also lower corn production in other states. Moreover, the farm crisis once again underlined the importance of a diversified farm operation, where livestock provides a farm family with income throughout the year.

But even though the pollution of ground water is a serious problem in Iowa and the corn belt, it pales beside the situation found in the Westlands Water District of California, where environmental degradation and excess production are found side by side. In the next few years federal and state authorities and the ranchers themselves will face the most difficult of decisions. They will have to determine whether it is possible to continue operations in an environment where the costs of doing business could rise dramatically. This process will strain the legal, political, economic, and social resources of the area to the limit.

The argument that the medium-sized family farm can survive and make a sizable contribution in the future rests partly on the historical evidence assembled here. This book, to some extent, is a celebration of the virtues that have allowed farm families and their communities to thrive in the corn belt and in the Central Valley. Studying the southern San Joaquin has allowed a firsthand appraisal of the super farm. If the evidence from the Tulare Lake bottom and elsewhere in the southern San Joaquin has any relevance, the super farm has already had a fair trial. In the boom years its performance was admirable, but in the downturn its organization was so bloated in land, machines, unsold commodities, and employees that it was more vulnerable than the medium-sized farm. Its operating budget of many millions of dollars could not be sustained by lenders looking for every opportunity to scale down their support of agriculture. In the volatile future, its limitations seem obvious.

Moreover, the farm crisis has alerted rural America, as never before, to the direct relationship between a depopulated countryside and large-scale farming. Although the Tulare Lake bottom was, and is, a unique environment, it has for many years acted as a laboratory in which to test the relationship between large-scale farming and the quality of community life. The results speak for themselves. Thus, despite the predictions for high technology and increased scale in agriculture, the evidence suggests otherwise. Assuredly the farm crisis has affected some of the medium-sized operations studied in the several communities covered here. However, the downturn has brought back into fashion some of the qualities that sustained such families in the past. If anything, the farm crisis made the "frugal farmer" mentality respectable once again.

The development of positive change rests largely on the alteration of attitudes and values. A more sustainable and diversified agriculture assuredly is possible. The work of the Small Farm Resources Project of the Center for Rural Affairs in Nebraska, is an example of what can be done by corn-belt farmers in this direction.[5] In addition, organizations such as the California Association of Family Farmers are dedicated to the promotion of similar ideals in the production of fruits and vegetables.[6] Thus, there are possibilities that in the not too distant future, not only the land but the farmers themselves can be rescued from the ravages of industrialized agriculture.

A Note on Sources and Methods

The social and economic history of agriculture is well endowed with primary and secondary sources. Because scientific research played a key role in farming, a tradition of collecting, manipulating, and processing agricultural data was established in the nineteenth century at the federal and state levels.

The voluminous United States Department of Agriculture bulletins and the agricultural census are just two examples of these kinds of materials. States, especially midwestern ones like Iowa, produced similar data and studies, and a certain percentage of this material dealt with economics and sociology. For example, Iowa conducted a state census every decade from 1875 to 1925, which reflected the passion for statistics that grew out of a tradition of scientific agriculture. Although used only sparingly in this study, both the 1915 and 1925 Iowa manuscript censuses were remarkable documents, which furnished data that was unavailable in the federal census until 1940. This study could not have been completed without the many detailed Iowa Agricultural Experiment Station bulletins dealing with the economic and social aspects of farming that were published from the twenties onwards.

For a number of reasons the California tradition was different. There was no state census, and primary data dealing specifically with farmers and ranchers, as opposed to farm labor, was collected and generated by federal agencies. The United States Immigration Commission's *Immigrant Farmers in the Western States* (Washington, D.C.: Government Printing Office, 1911), the Bureau of Agricultural Economics studies of the 1940s, and the La Follette hearings, for example, were valuable sources of information on the structure and organization of California agriculture. The California Experiment Station placed little emphasis on economic, social, and community research as it related to agriculture from the twenties to the sixties. However, the controversy over the role of the University of California in agricultural research in the late sixties and early seventies had an impact, and the contemporary period was served far better.

Ironically, though secondary sources were scanty in California compared to Iowa, primary sources in California courthouses were better organized and more readily available. In the southern San Joaquin the importance of operating loans ensured that such economic records as chattel mortgages were preserved. Similarly oil exploration and exploitation gave land records a greater visibility than was the case in a purely agricultural county.

Research that uses land and court records, especially when linkage is required, is both time consuming and costly. Therefore, any means of easing the burdens of research with these kinds of materials is desirable, indeed, a necessity. Indexes and tract books in private abstract and title companies were essential research tools. Therefore, counties were chosen for study not only if courthouse records were available and staff cooperative but also if it was possible to utilize tract books and indexes in the possession of private firms.

In California the choice of Kings County, and to a lesser extent Tulare, was made for these reasons. Arguably in the San Joaquin Valley, Fresno is the most representative agricultural county. At the same time, its very size and the fact that its courthouse is located in a large city, made such a choice impractical from a logistical point of view. In any case Kings provided a felicitous choice. The early history of its agriculture was representative of the San Joaquin as a whole. The experience on the lake bottom and the west side after 1945 was also representative of the extraordinary industrialization of agriculture that made the Central Valley so remarkable.

Much of the detailed archival work in Iowa took place in two counties: Fayette, in the northeastern part of the state, and Benton, in the east-central region. Again logistics and record availability played an important part in the selection process. Concern over "representativeness" was less critical in Iowa, for the large technical secondary literature permitted better coverage for those areas of the state where detailed research did not take place. Both counties had a tradition of livestock raising and also contained pockets of stable population, where farm families had deep roots in their neighborhoods and communities.

While much of the material for this study was based on conventional historical sources—newspapers, census materials, government reports, and the secondary literature—extensive interviewing and firsthand observation was required to explore the contemporary situation of farm families and the farm crisis.

Though the families that appear as cases here were not drawn from a scientific sample, many hours of research in the archives before going into the field, in addition to time spent with the families working on field tasks and sharing community and social activities, provided thorough knowledge of their background and present situation to ensure that they were typical of their particular milieu. Much of the material on the farm crisis in Iowa was gathered while I was engaged in a study of intergenerational relations in farm families. During a period of eighteen months, I attended many community meetings and conducted many interviews with farm advocates, lenders, clergy, social workers, as well as farmers themselves. I also made two short visits to California, in 1985 and 1986, after an extensive stay in 1984. Again, wherever possible, I interviewed key persons in banking, the Extension Service, agribusiness, and ranching, in order to obtain a thorough understanding of the complex situation the downturn created in the Central Valley.

Notes

Introduction

1. Quoted in James H. Shideler, *Farm Crisis, 1919- 1923* (Berkeley: University of California Press, 1957), 195.

2. The phrase is Judge Richard Stageman's, a bankruptcy judge of the Southern District of Iowa, see Des Moines *Register,* Nov. 7, 1985.

3. In Iowa 48.4 percent of farmers worked no days off the farm in 1982, in California 35.9 percent could be classified in the same category. In the San Joaquin Valley counties of Kings and Tulare 40.9 percent and 40.1 percent, respectively, worked no days off the farm. In the Iowa counties of Fayette and Benton, the ratios were 52.8 percent and 49.9 percent. Department of Commerce, Bureau of the Census, *1982 Census of Agriculture,* California, Iowa (Washington, D.C.: Government Printing Office, 1984), 162, 166, 192.

4. Donald Worster, "Hydraulic Society in California: An Ecological Interpretation," *Agricultural History* 56 (1982): 503-15.

5. Dean MacCannell and Edward Dolber-Smith, "Report on the Structure of Agriculture and Impacts of New Technologies on Rural Communities in Arizona, California and Texas," U.S. Congress, Office of Technology Assessment, *Technical Supplement to the Report on Technology, Public Policy and the Changing Structure of Agriculture: A Special Report on the 1985 Farm Bill, December 1985,* 9–18. For a treatment of modern industrialized agriculture and its implications in vegetable production, see William H. Friedland, Amy Barton, and Robert J. Thomas, *Manufacturing Green Gold: Capital, Labor and Technology in the Lettuce Industry* (New York: Cambridge University Press, 1981); and Robert J. Thomas, *Citizenship, Gender, and Work: Social Organization of Industrial Agriculture* (Berkeley: University of California Press, 1985).

6. Carey McWilliams, *California: The Great Exception* (New York: A.A. Wyn, 1949), 98-126.

7. Dan Walters, *The New California: Facing the 21st Century* (Sacramento: California Journal Press, 1986), 118.

8. B.F. Stanton, *Changes in Farm Structure: The United States and New York, 1930-1982,* Cornell Agricultural Economics Staff Paper 84-23 (1984), 1-2. For a similar definition of "small" California ranches, see Suzanne Vaupel, *Small Family Farms in California: The Definition Dilemma,* Small Farm Center, University of California at Davis, Farm Family Series (July 1986), 6.

9. Lawrence Goodwyn, *Democratic Promise: The Populist Movement in America* (New York: Oxford University Press, 1976), 25-423.

10. Willard W. Cochrane, *The Development of American Agriculture: A Historical Analysis* (Minneapolis: University of Minnesota Press, 1979), 116-21. For political insurgency in Iowa in the twenties, see Jerry Alvin Neprash, *The Brookhart Campaigns in Iowa, 1920-1926* (New York: Columbia University Press, 1932).

11. John L. Shover, *Cornbelt Rebellion: The Farmers' Holiday Association* (Urbana: University of Illinois Press, 1965), 114-31.

12. Cletus E. Daniel, *Bitter Harvest: A History of California Farmworkers, 1870-1941* (Ithaca: Cornell University Press, 1981), 167-221.

13. Cochrane, *Development of American Agriculture,* 140- 44.

14. Editorial in *California Farmer,* June 21, 1986, 4.

15. Donald Kaldor et al., *The Impact of New Industry on an Iowa Rural Community,* Iowa Agricultural Experiment Station (AES) Special Report 37 (1964), deals with rural industrialization in the 1950s and describes its effect on the lives of former and part-time farmers and their wives, who left farm jobs.

16. Ray E. Wakely, "How Changes in Population, Family, and Community Affect Agricultural Adjustment," in *A Basebook for Agricultural Adjustment in Iowa Agriculture in the Mid-Fifties,* Iowa AES Special Report 20 (1958), 34.

17. J. Tevere Macfadyen, *Gaining Ground: The Renewal of America's Small Farms* (New York: Ballantine, 1985), 111-12.

18. See Roe C. Black, "The New Era for Agriculture," *Farm Journal* (March 1977): 21, for a typically optimistic assessment of the farm situation in the middle to late seventies. See also Bryan Jones, *The Farming Game* (Lincoln: University of Nebraska Press, 1983), for a tongue-in-cheek description of farming in the "glory days" of the seventies and early eighties. Gilbert C. Fite, *American Farmers: The New Minority* (Bloomington: Indiana University Press, 1981), provides an insightful interpretation of agriculture just before the downturn.

19. *Economist* (London), Dec. 1, 1984, 31, 34.

20. Wendell Berry, *The Unsettling of America: Culture and Agriculture* (San Francisco: Sierra Club Books, 1977), takes a critical view of the practices of production agriculture. Frederick H. Buttel, "Agriculture, Environment, and Social Change: Some Emergent Issues," in *The Rural Sociology of Advanced Societies: Critical Perspectives,* ed. Buttel and H. Newby (Montclair, N.J.: Allanheld, Osmun, 1980), 467-72; Ingolf Vogeler, *The Myth of the Family Farm: Agribusiness Dominance of U.S. Agriculture* (Boulder, Colo.: Westview, 1981).

21. Robert G. Fellmeth, *The Politics of Land: Ralph Nader's Study Group Report on Land in California* (New York: Grossman, 1973).

22. Senate Select Committee on Small Business, *Will the Family Farm Survive in America?* Joint Hearings (Washington, D.C.: Government Printing Office, 1976), pt. 1C, provides documents collected by National Land for the People in their drive to break up large landholding on the west side of the San Joaquin; Don Villarjo, *New Lands for Agriculture: The California State Water Project* (Davis: California Institute for Rural Studies, 1981).

23. *The Family Farm in California,* Report of the Small Farm Viability Project (Sacramento: State of California, 1977).

24. Iowa Farmers Union, "Iowa Landownership Survey: A Preliminary Report on Land Tenure and Ownership in 47 Iowa Counties (Unpublished report, 1982). The Catholic bishop of Des Moines, the Reverend Maurice Dingman, an Iowa farm boy himself, was influential in speaking out on behalf of family farming. He was partly responsible for Pope John Paul's visit to Des Moines in 1979. See *Iowa Farmer Today,* Dec. 8, 1984.

25. USDA, *A Time to Choose: Summary Report on the Structure of Agriculture* (Washington D.C.: Government Printing Office, 1981).

26. David A. Stockman, *The Triumph of Politics* (New York: Harper and Row, 1986), 152-53, 154. Stockman's parents were Michigan fruit farmers who did not use farm programs.

27. Neil Harl, *Taxes and Agriculture,* Hearings before the Joint Economic Committee, 98th Cong., 2d Sess. (Washington, D.C.: Government Printing Office, 1985), 1049.

28. Neil Harl, "The Changing Rural Economy: Implications for Rural America" (Unpublished paper, Department of Economics, Iowa State University, 1985), 4–7; and Philip M. Raup, "What Prospective Changes May Mean for Agriculture and Rural

America," in *Farm Policy: The Emerging Agenda,* Missouri AES Special Report 338 (1985), 36, summarize the root causes of the farm crisis; ironically the importance of the Carter grain embargo in the downturn was questioned by a USDA study commissioned by Congress. See Des Moines *Register,* Nov. 29, 1986.

29. Shover, *Cornbelt Rebellion,* 18-21.

30. USDA, *Economic Indicators of the Farm Sector: State Income and Balance Sheet Statistics, 1983* (Washington, D.C.: Government Printing Office, 1984), 184, 196.

31. Quoted in Jonathan Rauch, "The Great Farm Gamble," *National Journal,* March 29, 1986, 759.

32. Fresno *Bee,* April 27, 1986.

33. William Schneider, "Opinion Outlook," *National Journal,* Dec. 21, 1985, 2930.

34. Ibid., 2931. See also Joseph J. Molnar and Patricia A. Duffy, "Urban and Suburban Residents' Perceptions of Farmers and Agriculture" (Unpublished paper, Department of Agricultural Economics and Rural Sociology, Auburn University, November 1986, kindly supplied by the authors).

35. For an application of this model to an agricultural setting, see J. Craig Jenkins, *The Politics of Insurgency: The Farm Worker Movement in the 1960s* (New York: Columbia University Press, 1985), 212.

36. Sockless Jerry Simpson of Kansas was an orator and organizer for the Farmers' Alliance movement of the 1880s and 1890s; Milo Reno a prominent figure in the Iowa Farm Holiday Association in the 1930s. See Goodwyn, *Democratic Promise,* 195–99; and Shover, *Cornbelt Rebellion,* 25-26.

37. Daniel, *Bitter Harvest,* 195-201.

38. William G. Murray and Willard O. Brown, *Farm Land and Debt Situation in Iowa,* Iowa AES Bulletin 328 (1935), 18. The forclosure figures were extrapolated from a sample of 16 counties; hence, the rounding of totals. Bankruptcy statistics can be found in David I. Wickens, *Farmer Bankruptcies, 1898-1935,* USDA Circular 414 (1936), 5-6. See Des Moines Register, Oct. 29, 1985, for the activities of the Dallas County, Iowa, Credit Council in the winter and spring of 1933. However, the mediation which took place in this locality was apparently unusual at that time in the Depression and a tribute to the "progressive" bankers and farmers living there.

39. For the influence of television on public and private behavior, particularly after the protests of the 1960s, see Joshua Meyrowitz, *No Sense of Place: The Impact of Electronic Media on Social Behavior* (New York: Oxford University Press, 1985), 308-9.

40. George Naylor, ed., *The United Farmer and Rancher Congress* (Ames, Iowa: North American Farm Alliance, 1986), 1.

41. For the experience of one member of the Farmers' Holiday Association who acted alone in northwest Iowa in 1932-33 without support of other activists, see Rodney D. Karr, "Farmer Rebels in Plymouth County, Iowa, 1932-1933," *Annals of Iowa* 47 (1985): 643-45.

42. These results are from the Iowa poll which included 334 farm households, Des Moines *Register,* Dec. 8, 1986.

43. Andrew H. Malcolm, *Final Harvest: An American Tragedy* (New York: Times Books, 1986).

Corn-Belt Farming

1. Lucy A. Stadley, *Relationship of the Farm Home to the Farm Business,* Minnesota AES Bulletin 279 (1931), 13-19.

2. Albert Mighell, *A Study of the Organization and Management of Dairy Farms in Northeast Ia.,* Iowa AES Bulletin 243 (1927), 56-61.

3. Ibid., 63-81.

4. Ibid., 82-87.

5. R.K. Buck, J.A. Hopkins, and C.C. Malone, *Dairy and Hog Farming in Northeast Iowa,* Iowa AES Bulletin 275 (1940), 685.

6. Ibid., 694-95, 706-7. See also R.K. Buck et al., *An Economic Study of the Dairy Enterprise in Northeastern Iowa,* Iowa AES Bulletin 277 (1940).

7. John A. Hopkins, Jr., *The Crop System of an Iowa County,* Iowa AES Bulletin 261 (1929), 286-88.

8. Ibid., 290-92.

9. John A. Hopkins, Jr., *An Economic Study of the Hog Enterprise in Humboldt County, Ia.,* Iowa AES Bulletin 255 (1928), 80-81, 92, 98.

10. John A. Hopkins, Jr., *Farm Organization and Management in Webster County, Ia.,* Iowa AES Bulletin 350 (1936), 393-404.

11. George A. Pond, *The Changing Pattern of Farming in Southeastern Minnesota,* Minnesota AES Bulletin 446 (1958), 9, 17-18. This study conveniently summarizes the revolution in agriculture in conditions similar to those of Iowa.

12. George M. Beal et al., *The Adoption of Two Farm Practices in a Central Iowa County,* Iowa AES Special Report 26 (1960), 7-12, 19.

13. M.A. Anderson et al., *An Appraisal of Factors Affecting the Acceptance and Use of Fertilizer in Ia. in 1953,* Iowa AES Special Report 16 (1956), 8-11.

14. H.O. Anderson et al., *The Economics of Some Soil Conservation Practices,* Iowa AES Research Bulletin 403 (1953), 579-608.

15. H.B. Howell, "Adjustments in Farm Size and Resources in Ia., Agriculture," in *A Basebook for Agricultural Adjustment in Iowa Agriculture in the Mid–Fifties,* Iowa AES Special Report 20 (1958), 31-33.

16. George W. Ladd, *Trends in the Iowa Dairy Industry,* Iowa AES Special Report 54 (1967), 14.

17. Eric O. Hoiberg and Wallace Huffman, *Profile of Iowa Farms and Farm Families, 1976* Iowa AES Bulletin 141 (1976), 26.

18. Ibid., 21-23.

19. Ibid., 29.

20. Iowa State University, *1984 Iowa Farm Costs and Returns,* Iowa Cooperative Extension Service Fm-1789 (1985), table 13.

21. Donald Johnson and Michael Boehlje, *Investment, Production, and Marketing Strategies for an Iowa Cattle Feeder in a Risky Environment,* Iowa AES Research Bulletin 592 (1981), 85-86.

22. For a summary of trends in the pork industry, see Roy N. Van Arsdall and Henry C. Gilliam, "Pork," in *Another Revolution in U.S. Farming,* ed. Lyle P. Schultz (Washington, D.C.: Government Printing Office, 1979), 236-48. See also Marty Strange and Chuck Hassebrook, *Take Hogs for Example: The Transformation of Hog Farms in America* (Walthill, Neb.: Center for Rural Affairs, 1981), for a critical assessment; Hoiberg and Huffman, *Profile of Iowa Farms,* 31- 33.

23. Van Arsdall and Gilliam, "Pork," 236-43.

24. Arnold Paulsen and Michael Rahm, *Development of Subsidiary Sow-Farrowing Firms in Iowa,* Iowa AES Special Report 83 (1979), 19.

25. ISU, *1984 Farm Costs and Returns,* tables 12, 14.

26. Ibid., table 14.

27. For a summary of structural changes in the dairy industry of the Upper Midwest, see Harry Kaiser and Jerome Howard, "The Changing Structure of the Minnesota Dairy Industry," in Department of Agricultural Economics, University of Minnesota, *Economic Report 83-8 (1983).* Ladd, *Trends in the Iowa Dairy Industry,* 13-14.

28. ISU, *1984 Farm Costs and Returns,* table 14.

Central Valley Ranching

1. J.L. Brown, *The Story of Kings County, California* (Berkeley, Calif.: Lederer, 1940), 80-81.

2. Frank Norris, *The Octopus: A Story of California* (New York: Doubleday, 1903).

3. R.L. Adams, and L.A. Crawford, *Tests of Farm Organization in the Turlock Area,* California AES Bulletin 544 (1932), 95-109.

4. R.L. Adams, *Cost of Producing Market Milk and Butter Fat,* California AES Bulletin 372 (1923), 73-83.

5. Adams and Crawford, *Tests of Farm Organization,* 50- 54.

6. Ibid., 55-57.

7. A.W. Christie and L.C. Barnard, *The Principles and Practice of Sun-Drying Fruit,* California AES Bulletin 388 (1925), 40-42, 48-54.

8. Ibid., 43, 44, 46-47.

9. Donald J. Pisani, *From the Family Farm to Agribusiness: The Irrigation Crusade in California and the West, 1850-1931* (Berkeley: University of California Press, 1984), 440-41.

10. California Department of Engineering, *Use of Water from Kings River, California* (Sacramento: State Printer, 1920).

11. Arthur Maass and Raymond L. Anderson, *...And the Desert Shall Rejoice: Conflict, Growth, and Justice in Arid Environments* (Cambridge: Massachusetts Institute of Technology Press, 1978), 177-96.

12. California Department of Engineering, *Use of Water from Kings River,* 28.

13. Ibid., 36-41.

14. Ibid., 44, 114-18.

15. Walter Camp, *Cotton Culture in the San Joaquin Valley,* USDA Circular 164 (1921); David C. Large, "Cotton in the San Joaquin Valley: A Study of Government in Agriculture," *Geographical Review* 48 (1958): 367, 369.

16. Howard Gregor, "The Regional Primacy of San Joaquin Valley Agriculture," *Journal of Geography* 61 (1962): 394-99.

17. Quoted in John Turner, *White Gold Comes to California,* (Bakersfield: California Cottonseed Distributors, 1981), 26. For reclamation, see Steven Zimrick, "The Changing Order of Agriculture in the Southern San Joaquin" (Ph.D. diss., Louisiana State University, 1976), 196-97.

18. George Peterson, Jr., *Cotton Production in the Southern Desert Valleys of California,* California AES Ciruclar 528 (1959), provides technical information on cotton culture for an area similar to the San Joaquin. For early irrigation techniques, see S.H. Beckett and Carroll F. Dunshee, *Water Requirements for Cotton in Sandy Loam Soils in the Southern San Joaquin,* California AES Bulletin 537 (1932); and Frank Adams, *Cotton Irrigation Investigations in the San Joaquin,* California AES Bulletin 668 (1938).

19. Interview with Frances Von Glahn, Corcoran, Calif., July 7, 1984.

20. Frank J. Taylor, "Farming Round the Clock," *Country Gentleman* 109 (1943): 10, 24.

21. Ibid., 24; Fresno *Bee,* July 1, 1948.

22. Trimble R. Hedges and Warren R. Bailey, *The Economics of Mechanical Cotton Harvesting,* California AES Bulletin 743 (1954), 4, 39.

23. William C. Beatty, Jr., "A Preliminary Report on a Study of Farm Laborers in Fresno County from Jan. 1, 1959 to July 1, 1959" (Fresno County Rural Health and Education Committee, 1959, mimeo), deals with the problem of underemployment as a result of structural change in agriculture. For material on labor contractors and their role, see 36-38; the bleak postwar period is covered in Ernesto Galarza, *Farm Workers and Agribusiness in California, 1947-1960* (Notre Dame, Ind.: Unviersity of Notre Dame Press, 1977). For a sophisticated analysis of the impact of the cotton harvester on labor in the South, see Willis Peterson and Yoav Kislev, "The Cotton Harvester in Retrospect: Labor Displacement or Replacement?" *Journal of Economic History* 46 (March 1986): 199-215.

24. Quoted in William G. Jeffs, "The Roots of the Delano Grape Strike" (M.A. thesis, California State University, Fullerton, 1969), 72, 23.

25. Robert DeRoos, *The Thirsty Land: The Story of the Central Valley Project* (Palo Alto, Calif.: Stanford University Press, 1948); Donald Worster, *Rivers of Empire: Water Aridity and the Growth of the American West* (New York: Pantheon, 1985), 244-56.

26. Maass and Anderson, *...And the Desert Shall Rejoice*, 243-45, 256-71; Marc Reisner, *Cadillac Desert: The American West and Its Disappearing Water* (New York: Viking, 1986), 183.

27. Ellen Liebman, *California Farmland: A History of Large Landholding* (Totowa, N.J.: Rowman and Allanheld, 1983), 154-55.

28. Scott Matulich et al., *Cost-Size Relationships for Large-Scale Dairies with Emphasis on Waste Management*, Giannini Foundation Research Report 324 (1977), is an exhaustive study of dairy waste management in the Chino Basin, where urban sprawl from Los Angeles, forced the removal of a number of dairies to the southern San Joaquin.

29. Arthur Shultis, *Dairy Management in California*, California AES Bulletin 640 (1940); J.M. Finley, *The Dairy Situation in Ca.*, California AES Circular 369 (1944); Arthur Shultis et al., *California Dairy Farm Management*, California AES Circular 417 (1963), 3, 4, 10.

30. Shultis et al., *Calif. Dairy Management*, 25.

31. Kings County Dairy Herd Improvement Association, *Annual Report, 1982* (Hanford, Calif.: Kings County DHIA, 1983), 45.

32. See Corcoran *Journal*, July 14, 1983; Hanford *Sentinel*, May 29, 1985.

33. This material can be found in Plaintiff's Response to Report of Appraisers from Shepard Farms, Inc., 27-30, in Probate Docket 11183, Kings County Clerk of Court, Hanford, Calif.

34. Ibid., 32-33.

35. Ibid., 34-35.

Land Tenure

1. For a critical appraisal of United States land tenure, see Leonard Salter, Jr., *A Critical Review of Research in Land Economics*, (Madison: University of Wisconsin Press, 1967), 5-38; and more recently, Jack R. Kloppenburg, Jr., and Charles C. Geisler, "The Agricultural Ladder: Agrarian Ideology and the Changing Structure of U.S. Agriculture," *Journal of Rural Studies* 1 (1985): 59-72.

2. The classic historical statement is, of course, found in Frederick Jackson Turner, "The Significance of the Frontier in American History," American Historical Association, *Annual Report for 1893* (Washington, D.C.: AHA, 1893), 199-227.

3. Richard T. Ely, "Landed Property as an Economic Concept and as a Field of Research, "*American Economic Review*, suppl. 7 (1917): 18-35.

4. Leonard Salter, Jr., *Land Tenure in Process*, Wisconsin AES Bulletin 146 (1943), 34.

5. For thorough discussion of this process in a corn-belt setting, see ibid., 35-38.

6. See the listings for Iowa and other midwestern states compared to those of California, in USDA, Bureau of Agricultural Economics, *Publications Relating to Rural Life, Issued by Various State Colleges of Agriculture* (Washington, D.C.: USDA, n.d.). The sole California publications were R.L. Adams and William H. Smith, Jr., *Farm Tenancy in California and Methods of Leasing*, University of California Experimnt Station Bulletin 655 (1940), which superseded Adams's Bulletin 272 (1923).

7. Walter Goldschmidt, *As You Sow* (Glencoe, Ill.: Free Press, 1947).

8. R.L. Adams, *California Farms: To Buy or Not to Buy?* California AES Circular 358 (1944), 3.

9. L.C. Gray, "Farm Ownership and Tenancy," *1923 Year Book of Agriculture* (Washington D.C.: USDA, 1924), 555.

10. Robert A. Rohwer, *Family Factors in Tenure Experience: Hamilton County, Iowa*, Iowa AES Bulletin 375 (1950), 856.

11. For a more elegant typology of this theme, see Sonja Salamon, "Ethnic Communities and the Structure of Agriculture," *Rural Sociology* 50 (1985): 325.

12. John F. Timmons, *Improving Farm Rental Arrangements in Iowa,* Iowa AES Bulletin 393 (1953), 82-83.

13. William L. Preston, *Vanishing Landscapes: Land and Life in the Tulare Lake Basin* (Berkeley: University of California Press, 1981), 121-89.

14. A useful summary of the development of the Kings River delta is found in Maass and Anderson, *...And the Desert Shall Rejoice,* 157-97.

15. Interview with Dale Kleinesser, Fresno, Calif., July 7, 1984.

16. Alvin Ray Graves, "Immigrants in Agriculture: The Portuguese Californians, 1850-1970's" (Ph.D. diss., University of California, Los Angeles, 1977), chaps. 1-6.

17. Deed Book 69: 346, 71:95, Fayette County Recorder's Office, West Union, Iowa.

18. Rainer Schickele and Charles A. Norman, *Farm Tenure in Iowa,* Iowa Agricultural Experiment Station Bulletin 354 (1937), 173.

19. Goldschmidt, *As You Sow,* 9.

20. See Keiko Hachimara, "Japanese Farming in Fresno County, Ca.: Prewar and Postwar Patterns" (M.A. thesis, California State University, Fresno, 1981), 29-30, for leasing practices between Anglo owners who were relieved of management responsibilities, labor hiring, etc. Karen Leonard, "Punjabi Farmers and California's Alien Land Law," *Agricultural History* 59 (Oct. 1985): 549-62, discusses the attitudes of Anglo landlords and discrimination towards their tenants in the similar environment of the Imperial Valley in this period. Sucheng Chan, *This Bitter-Sweet Soil: The Chinese in California Agriculture, 1860-1910* (Berkeley: University of California Press, 1986), is the definitive work on the Chinese.

21. The following cases were randomly selected from the lease books between 1900 and 1920, to reflect patterns in Kings County in that period. Lease Book 1:385, Dec. 8, 1900, Kings County Recorder's Office, Hanford, Calif.

22. Lease Book 2:4, Jan. 22, 1904, ibid.

23. Lease Book 2:329, Jan. 10, 1911, ibid.

24. Lease Book 2:283, April 5, 1910, ibid.

25. Lease Book 2:274, Dec. 29, 1909, ibid.

26. Lease Book 3:146, June 1, 1914, ibid.

27. For farmer retrenchment in the 1920 farm crisis, see Shideler, *Farm Crisis,* 76-83.

28. William G. Murray, *Farm Mortgage Foreclosures in Southern Iowa, 1915-1936,* Iowa AES Bulletin 248 (1938), 251, 257.

29. William G. Murray, *Corporate Land, Foreclosure, Mortgage Debt, and Land Values in Iowa, 1939,* Iowa AES Bulletin 266 (1939), 309.

30. Ibid., 314.

31. John Warren Thomas, "Factors Related to the Success and Failure of Farm Operators in Acquiring Farm Ownership in Milford Township, Story County, Ia., Since 1925" (M.A. thesis, Iowa State College, 1951), 143-81.

32. E.A. Schuler, *Social Status and Farm Tenure: Attitudes and Social Conditions of Cornbelt and Cotton Belt Farmers,* Bureau of Agricultural Economics, Social Report 4 (Washington D.C.: BAE, 1938), 165.

33. Robert Diller, *Farm Ownership, Tenancy, and Land Use in a Nebraska Community* (Chicago: University of Chicago Press, 1941), 47.

34. Rainer Schickele, *Farm Tenure in Iowa: Facts on the Farm Tenure Situation,* Iowa AES Bulletin 356 (1937), 245.

35. Theodore W. Schultz, "What Has Happened to the Agricultural Ladder?" in *Farm Tenure in Iowa: National Farm Institute Symposium,* Iowa AES Bulletin 357 (1937).

36. Schickele and Norman, *Farm Tenure in Iowa,* 166, 170-72.

37. Clarke A. Chambers, *California Farm Organizations* (Berkeley: University of California Press, 1952), 39-52, 70-81; Senate Committee on Education and Labor, *Hearings before a Subcommittee to Investigate Violations of the Right of Free Speech and Assembly and Interference with the Right of Labor to Organize and Bargain Collectively*, 74th Cong., 3d Sess. (Washington, D.C.: Government Printing Office, 1940), 26424-29.

38. R.L. Adams, *California Farms*, 3.

39. Varden Fuller, "The Supply of Agricultural Labor as a Factor in the Evolution of Farm Organization in California" (Ph.D. diss., University of California, Berkeley, 1939).

40. Interview with James Stout, Hanford Calif., July 4, 1984.

41. State Relief Administration of California, "A Modern Patriarch," in "migratory Labor in California" (California Relief Administration, 1936, mimeographed), 133-36.

42. Ibid., 139. Steinbeck's sources for the novel are discussed in Jackson J. Benson, *The True Adventures of John Steinbeck, Writer* (New York: Viking, 1984), 302-9; however, no mention is made of Clay Bennett.

43. State Relief Administration, "A Modern Patriarch," 137-38.

44. "The Effect of the Central Valley Project on Agricultural and Industrial Economy and the Social Character of California," Report on Problem 24, Central Valley Project Studies (Bureau of Agricultural Economics, 1945, mimeographed), 100.

45. Bureau of Agricultural Economics, "Rural Trends in Tulare County," 78, Records of the Regional Representative of the Division of Farm Population and Rural Welfare, Box 20, RG 83, National Archives, San Bruno, Calif.

46. "The Effect of the Central Valley Project," 84.

47. William H. Smith, "Variations in Farm Organization Associated with the Tenure of Cotton Farms in Tulare County, 1940" (M.A. thesis, University of California, Berkeley, 1940), 19, 25-26.

48. Ibid., 27-32.

49. BAE, "Rural Trends in Tulare County," 74.

50. See, for example, H.B. Howell, "Adjustments in Farm Size and Resources."

51. This classification comes from Richard P. Rodefeld, "Trends in U.S. Farm Organizational Structure and Type," in *Change in Rural America*, ed., Rodefeld et al. (St. Louis: C.V. Mosby, 1978), 158-77.

52. John F. Timmons and Raleigh Barlowe, *Farm Ownership in the Midwest*, Iowa AES Bulletin 361 (1949), 857.

53. Bureau of Census, *Agriculture* (Washington, D.C.: Bureau of Census, 1951), 58.

54. Timmons and Barlowe, *Farm Ownership in the Midwest*, 904-5.

55. Roger W. Strohbehn and John F. Timmons, *The Ownership of Iowa's Farmland*, Iowa AES Bulletin 489 (1960), 20.

56. Ibid., tables 16, 17, p. 14.

57. Timmons, *Improving Farm Rental Arrangements*, 96-98.

58. E.B. Hill, *Father and Son Partnerships*, Michigan AES Bulletin 330 (1944), 20-28.

59. Robert Rohwer, *The Social Status and Occupational Prospects of Married Farm Laborers in Cherokee County, Ia.*, Iowa AES Bulletin 452 (1957), 120.

60. Donald R. Kaldor, *Characteristics of Operator Entry into Iowa Farming, 1959-60*, Iowa AES Bulletin 546 (1964), 754-759.

61. Ibid., 761.

62. Randall H. Hoffman and Earl O. Heady, *Production Income and Resource Changes from Farm Consolidations*, Iowa AES Bulletin 502 (1962), 382.

63. Donald R. Kaldor and William M. Edwards, *Occupational Adjustment of Iowa Farm Operators Who Quit Farming in 1959-61*, Iowa AES Special Report 75 (1975), 3-7.

64. Worster, *Rivers of Empire*, 241-56. For tenure in the southern San Joaquin before construction of the Central Valley Project, see Edwin E. Wilson and Marion Clawson, "Agriculture, Landownership, and Operations in the Southern San Joaquin" (Berkeley, Calif.: BAE, 1945, mimeographed).

65. *Family Farm in California*, 6-17.

66. Goldschmidt, *As You Sow*, 22.

67. W.E. Johnston, *Some Characteristics of the Farm Real Estate Market in California with Emphasis on Transactions in Imperial and Tulare Counties*, California AES Bulletin 856 (1971), 36-37.

68. The Chicano cooperative movement provides an exception to this generalization, but apparently there was only one cooperative in the southern Central Valley, at Yettem in Tulare County. See Refugio I. Rochin, "The Conversion of Chicano Farm Workers into Owner-Operators of Cooperative Farms, 1970-1985," *Rural Sociology* 51 (1986): 97-115.

69. Between 1960 and 1969, 46.6 percent of all farming corporations in Calif. were formed. Most of these had ten or fewer stockholders. See C.V. Moore and J.H. Snyder, *A Statistical Profile of Ca. Corporate Farms*, Giannini Foundation Information Series in Agricultural Economics 70-3 (1970), 15.

70. Yoshino Tajiri Hasegawa and Keith Boettcher, eds., "Success Through Perseverence: Japanese Americans in the San Joaquin Valley," Japanese-American Project (Fresno: San Joaquin Valley Library System, 1980, mimeographed), 37-55; and Hachimara, "Japanese Farming in Fresno County," 70-84.

71. To preserve anonymity all names have been changed. For the organization of the Payne Company, see Steven Zimrick, "The Changing Organization of Agriculture in the Southern San Joaquin Valley, Ca." (Ph.D., diss., Louisiana State University, 1976), 200-202. For an excellent analysis of the Payne Company farming strategy in the sixties and early seventies, see Pamela Joy Merrill, "American Involvement and the Resurgence in the Australian Cotton-Growing Industry, 1962-72" (Ph.D. diss., University of California, Berkeley, 1977), 140-62.

72. Robert E. Leake, Jr., and Stanley M. Barnes, *The Pine Flat Project on the Kings River, Ca.—Reclamation Law Should Not Apply! A Briefing Paper on the Issue* (Fresno: Kings River Water Users Committee, 1980), 16.

73. For an analysis of agricultural conditions on the west side during the era of deep-well pumping, see Chester F. Cole, "Rural Occupance Patterns in the Great Valley Portion of Fresno County, Ca." (Ph.D. diss., University of Nebraska, 1950), 483-509.

74. Ed Simmons, *Westlands Water District: The First 25 Years* (Fresno: Westlands Water District, 1983), 32-75, provides background on the history of federal intervention on the Westside.

75. Senate Select Committee on Small Business, *Will the Family Farm Survive?*

76. Liebman, *California Farmland*, 156-57.

77. Worster, *Rivers of Empire*, chap. 6. See also E. Philip LeVeen, "Enforcing the Reclamation Act and Rural Development in Ca., *Rural Sociology* 44 (1979): 667-70, which found that a hypothetical breakup of farms into 320-acre units would not be beneficial to "small farmers" but was more likely to strengthen the position of medium-sized operations; the average farm size in Westlands was 1,654 acres, with a median of 690 acres. 5.2 percent were corporations with more than ten stockholders; 32.7 percent were incorporated with fewer than ten stockholders; 35.4 percent were partnerships with family members; 15 percent were partnerships with spouses only; 10.1 percent were owned by individuals; and 1.3 percent were trusts or estates. See Charles V. Moore et al., *Structure and Performance of Western Irrigated Agriculture*, Giannini Foundation Information Series Bulletin 1905 (1982), 20; telephone interview with Michael Heaton, assistant general counsel for Westlands, Jan. 5, 1987.

78. *Washington Post National Weekly Edition*, 25 May 1987; *Fresno Bee*, 12 April 1987.

Inheritance

1. For a theoretical discussion of the dynamics of the farm family, see John W. Bennett, *Of Time and the Enterprise: North American Farm Management in a Context of Resource Marginality* (Minneapolis: University of Minnesota Press, 1982), 113; and Sonja Salamon and Shirley M. O'Reilly, "Family Land and Development Cycles among Illinois Farmers," *Rural Sociology* 44 (1979): 525-42. For the historical period, see Kathleen Neils Conzen, "Peasant Pioneers," in *The Countryside in the Age of Capitalist Transformation*, ed. Steven Hahn and Jonathan Prude (Chapel Hill: University of North Carolina Press, 1985), 259-92.

2. Kenneth H. Parsons and Eliot O. Waples, *Keeping the Farm in the Family*, Wisconsin AES Bulletin 157 (1945); and Eliot O. Waples, "Farm Ownership Processes in a Low Tenancy Area" (Ph.D. diss., University of Wisconsin, 1946).

3. Parsons and Waples, *Keeping the Farm in the Family*, 6; W.H. Spiegel, "The Altenteil: German Farmers' Old Age Security," *Rural Sociology* 4 (1939): 201-18; Carl F. Wehrwein, "Bonds of Maintenance as Aids in Acquiring Farm Ownership," *Journal of Land and Public Utility Economics* 8 (1932): 396-403.

4. Diller, *Farm Ownership, Tenancy, and Land Use*, 134.

5. For a summary of community property law in the historical context, see Chester Varnier, *American Family Law: Husbands and Wives*, 5 vols. (Palo Alto: Stanford University Press, 1935), 3:207-54.

6. Russell L. Berry and Elton B. Hall, *How to Keep Your Farm in the Family*, Michigan AES Bulletin 357 (1949); and John F. Timmons and John C. O'Byrne, *Transferring Farm Property within Families in Iowa*, Iowa AES Research Bulletin 394 (1953).

7. Varnier, *American Family Law*, 3:281-321.

8. For Central Europe, see Herman Rebel, *The Bureaucratization of Property and Family Relations under Early Hapsburg Absolutism, 1511-1636* (Princeton: Princeton University Press, 1983), 170-229; and for Norway, see John Gjerde, *From Peasants to Farmers* (New York: Cambridge University Press, 1985), 85-87.

9. Gunter Golde, *Catholics and Protestants: Agricultural Modernization in Two German Villages* (New York: Academic Press, 1975), 76.

10. Parsons and Waples, *Keeping the Farm in the Family*, 27- 28.

11. Timmons and O'Byrne, *Transfering Farm Property*, 176- 80.

12. Ibid., 170-76.

13. Samuel F. Pearce and Harry Low, *Estate Planning for Farmers*, California AES Circular 461 (1957), 1-6.

14. Theoretically one could draw a random sample of Iowa farmers in 1900 from the manuscripts of the United States Census, and trace through tax and probate records to build a composite picture of land tenure and inheritance. Unfortunately in many Iowa counties tax records have been destroyed, and time and distance made such a research design all but impossible. Instead, ten townships across the state were selected because the counties in which they were located had the necessary records available and abstract companies cooperated by permitting searches in their records. For a discussion of the difficulties of rural probate and land record searches, see Mark Friedberger, "Probate and Land Records in Rural Areas," *Agricultural History* 58 (1984): 123-26.

15. Marcus Lee Hansen, *The Immigrant in American History* (Cambridge: Harvard University Press, 1940), 61-62.

16. Deed Book 104:7, Fayette County, Recorder's Office, West Union, Iowa.

17. Deed Book 60:362, ibid.

18. Horace Miner, *Culture and Agriculture: An Anthropological Study of a Corn Belt County* (Ann Arbor: University of Michigan Press, 1949), 27.

19. Probate Docket, 5342, Benton County Clerk of Court, Vinton, Iowa.

20. Probate Docket 7764, ibid.

21. The multiple regression analysis predicted which variables had the greatest effect on long-term landownership. The following equation isolated the significant variables (with standard errors in parentheses):

$$\text{Farm in family} = -.324 + \underset{(.065)}{328\ \text{intervivos}} + \underset{(.002)}{.008\ \text{age}} + \underset{(.052)}{.191\ \text{German}} + \underset{(.014)}{.036\ \text{males}}$$

$$R2 = .18 \qquad N = 332.$$

All four independent variables were significant at better than the .01 level. Standard regression coefficients were .26 for intervivos transfer, .19 for age of transfer and ethnicity, and .13 for the number of male heirs. Age of transfer, though a better predictor than ethnicity or male heirs, had little quantifiable or interpretable effect. An intervivos transfer gave a family a 32 percent better chance of keeping the farm for two or more generations. Those with German background had a 19 percent better chance of being on the same land two generations later, and any male heir increased the chances by 3 percent.

22. I drew a random number of probate cases were drawn from the Kings County dockets, 1900-55, to reflect changes in inheritance behavior. Probate Docket 524, Feb. 6, 1912, Kings County Clerk of Court, Hanford, Calif.

23. Probate Dockets 199, 212, May 14, June 4, 1901, ibid.

24. Probate Docket 514, Sept. 21, 1910, ibid.

25. Probate Dockets 1577, 1324, Dec. 17, 1925, Nov. 3, 1922, ibid.

26. Probate Docket 547, March 22, 1911, ibid.

27. Probate Docket 792, Dec. 27, 1916, ibid.

28. Probate Dockets 2304, 2311, Jan. 7, 27, 1936, ibid.

29. Probate Docket 2021, Feb. 13, 1931, ibid.

30. Probate Docket, 6422, May 16, 1954, ibid.

31. Probate Docket 3976, March 24, 1948, ibid.

32. Probate Docket 3867, Sept. 18, 1947, ibid.

33. Ward W. Bauder, *Iowa Farm Operator's and Landlords' Knowledge of, Participation in and Acceptance of the Old Age and Survivors Insurance Program*, Iowa AES Bulletin 489 (1960), 22.

34. W. Fred Woods, *Increasing Impact of Federal Estate and Gift Taxes on the Farm Sector,* USDA Agricultural Economic Report (1973), 242.

35. Buel Franklin Lampher, Jr., "Problems and Implications of Intra-family Farm Property Transfers in Grundy County, Ia." (Ph.D. diss., Iowa State College, 1955), 244-45.

36. Strohbehn and Timmons, *Ownership of Iowa's Farmland*, 22.

37. Lampher, "Intra-family Farm Transfers," 41.

38. Woods, *Increasing Impact of Estate and Gift Taxes*, 6.

39. Michael Boehlje et al., *Intergenerational Transfer and Estate Planning: The Iowa Experience*, Iowa AES Special Report 84 (1979), 9.

40. Ibid., 16.

41. Charles Davenport and Michael Boehlje, *The Effects of Tax Policy on American Agriculture,* USDA Agricultural Economic Report 480 (1982), 10, 18-21.

42. S.L. Titus, P.C. Rosenblatt, and R.M. Anderson, "Family Conflict over Inheritance of Property," *Family Coordinator* 28 (1979): 337-46. For the impact of farm women in the business, see Des Moines *Sunday Register,* April 8, 1979.

43. Los Angeles *Times,* March 27, 1983.

44. See, for example, Civil Cases 5819, 5846, 7776, 9740, which deal with disputes over unpaid bills, trespass, and the planting and harvesting of a crop on the land of another party, Kings County Clerk of Court.

45. See disposition in Civil Case 24807, Kings County Clerk of Court.

46. See Los Angeles *Times*, March 27, 1983; Fresno *Bee*, April 12, 1983.

47. These Memoranda are filed in *Wood Land Co., et al.* v. *Jane Wood et al.*, Civil Case 36007, Kings County Clerk of Court, hereinafter cited as Memoranda.

48. Memoranda, 6-10.

49. Los Angeles *Times*, March 27, 1983.

50. Memoranda, 19-23.

51. This handwritten letter is found in Civil Case 36427, *Wood Land Co.* v. *Wood*, Kings County Clerk of Court.

52. Los Angeles *Times*, June 18, 1982.

53. See Civil Case 38895, Kings County Clerk of Court.

54. Hanford *Sentinel*, May 28, 1985.

55. See Probate Docket 11878, Kings County Clerk of Court.

56. Probate Docket 11183, ibid.

57. See depositions contained in Probate Docket 11515, ibid.

58. Interview with John Shepard, Orosi, Calif., Dec. 12, 1984; Civil Case 34889, *Parrish* v. *Shepard*, Kings County Clerk of Court.

Credit

1. For a pioneering historical study of mortgage borrowing for farmers, see Alan G. Bogue, *Money at Interest* (Ithaca: Cornell University Press, 1955); and also his "Land Credit for Northern Farmers, 1789-1940," *Agricultural History* 50 (Jan. 1976): 68-100. For a useful introduction to the theoretical background to the credit needs of farm families in contemporary developing countries, see Dale W. Adams, "Effect of Finance on Rural Development," in *Undermining Rural Development with Cheap Credit*, ed. Adams et al. (Boulder, Colo.: Westview, 1984), 12-16.

2. Lowell K. Dyson, "Was Agricultural Distress in the 1930s a Result of Land Speculation during World War I? The Case of Iowa" *Annals of Iowa* 34 (1971): 577-84.

3. William G. Murray, *An Economic Analysis of Farm Mortgages in Story County, Ia., 1854-1931*, Iowa AES Bulletin 156 (1933), 368-75.

4. Ibid., 377-79.

5. Ibid., 385-93.

6. Ibid., 395-99.

7. Murray, *Farm Mortgage Foreclosures in Southern Iowa*, 252.

8. Dyson, "Was Agricultural Distress," 578.

9. L.C. Gray and O.G. Lloyd, *Farm Land Values in Iowa*, USDA Bulletin 874 (1920), 12-14.

10. Ibid., 39-41.

11. Quoted in Dyson, "Was Agricultural Distress," 578.

12. For bank failures in the twenties as a result of the downturn, see Fred L. Garlock, *Long Term Loans of Iowa Banks*, Iowa AES Research Bulletin 129 (1930), 298-300.

13. See Deed Book 51:185, 61:323, 363, 66:430, Fayette County Recorder's Office, West Union, Iowa, for land sales and mortgage arrangements with the Zeigler Company.

14. Deed Book 64:53-54, ibid.

15. Interview with Nick Kuennen, Auburn, Iowa, Jan. 4, 1983; B.H. Hibbard and Frank Robitka, *Farm Credit in Wisconsin*, Wisconsin AES Bulletin 247 (1915), 44-52.

16. Deed Book 99:456, and see 147:457, 148:156, Fayette County Recorder's Office, for successful transactions of high-priced land in the 1920s.

17. See William Dobbert to Aetna Life, March 1894, William Wirth to Burlington Savings, June 1900, Michael Kearney to Aetna Life, April 1900, Township 91 North, 45 West, Plymouth County Abstract Company Tract Book, LeMars, Iowa.

18. Maurice Fred Perkins, "Farm Mortgage Commissions in North Central Iowa, 1910-25" (M.A. thesis, Iowa State College, 1940), 29-33.
19. Hugh H. Shepard to L.H. Bush, May 1, 1925, in Nelson File 1247, Shepard Papers, Iowa State University Archives, Ames.
20. Hugh H. Shepard to Kate Dunlay, February 21 1930, in Dunlay File 2039, Shepard Papers.
21. Roy J. Barrows, *Experience of Michigan Rural Banks with Short Term Loans to Farmers*, Michigan AES Special Bulletin 311 (1941).
22. Ibid., 12-23.
23. Ibid., 35-40.
24. Ibid., 41-43.
25. Tract Book, Section 7, Township 18 South, R22 East, Kings County Title Company, Hanford, Calif.
26. Tract Book, Section 21, Township 19 South, R22 East, ibid.
27. Tract Book, Section 23, Township 16 South, R23 East, Safeco Title Company, Visalia, Calif.
28. Tract Book, Section 30, Township 18 South, R27 East, ibid.
29. Mortgages were chosen from pages of the official records in 1930 and 1946. Official Records, Book 67: 136, Oct. 30, 1930, Kings County Recorder's Office, Hanford, Calif.
30. Official Records, Book 66:467, 480, Oct. 19, 20, 1930, Book 67: 248, Nov. 24, 1930, ibid.
31. Official Records, Book 341:43, Jan. 16, 1946, Book 344:88, Feb. 23, 1946, Book 367:63, Dec. 26, 1946, Book 342:101, Jan. 29, 1946, ibid.
32. For the diffusion of cotton in the 1920s, see John Turner, *White Gold Comes to California*, an "in house history" but reliable on early facts.
33. The contribution of the ginning companies to the financial support of cotton is treated with breathless enthusiasm in "California's Cotton Rush," *Fortune* (1949): 136-37.
34. W. Clifford Hoag, *The Farm Credit System: A History of Financial Self Help* (Danville, Ill.: Interstate, 1979), 219-73.
35. Aaron Gustave Nelson, "Experience of the Federal Land Bank with Loans in Four North Central Iowa Counties, 1917-47" (Ph.D. diss., Iowa State College, 1949), 25.
36. Ibid., 22-24.
37. Ibid., 49-52.
38. Robert W. Wilcox, "The Farmer's Home Administration Farm Ownership Program in Iowa" (Ph.D. diss., Iowa State College, 1947), 40.
39. Ibid., 65-77, discusses individual cases.
40. Sherwood O. Berg et al., *Loans of Production Credit Associations to Minnesota Farmers*, Minnesota AES Bulletin 410 (1952), 10-18, 27-28.
41. For conservative borrowing behavior, see Gordon G. Bevins, *Use of Credit by Families in Southern Iowa and Northern Missouri*, Iowa AES Special Report 35 (1963), 11-13.
42. Cooperative Farm Credit System, *34th Annual Report, 1966-67* (Washington, D.C.: Government Printing Office, 1968), 20, 63.
43. *Family Farm in California*, 73-80.
44. Willard W. Cochrane and Mary E. Ryan, *American Farm Policy, 1948-1973* (Minneapolis: University of Minnesota Press, 1976), 383-91.
45. Cochrane, *Development of American Agriculture*, 144-47.
46. Geoffrey Shepherd, *Appraisal of Federal Feed Grain Programs*, Iowa AES Research Bulletin 501 (1962), 354-57.
47. USDA, *Economic Indicators of the Farm Sector: State Income and Balance Sheet 1984* (Washington: Government Printing Office, 1985), 74; Cochrane and Ryan, *American Farm Policy*, 366.

48. Cochrane and Ryan, *American Farm Policy,* 366; Ladd, *Trends in the Iowa Dairy Industry,* 13.

49. USDA *Economic Indicators of the Farm Sector,* 168, 174.

50. Des Moines *Sunday Register,* July 6, Dec. 28, 1986.

51. Fresno *Bee,* April 27, 1986; Fresno *Bee,* June 7, 1987.

52. Cornelius Gallagher, "Financial Aspects of a Farm Operation from a Banker's Perspective" (Unpublished speech, Bank of America, 1981), 1-3; and Cornelius Gallagher, "Debt Control Vital to Business Survival," *California Farm Press,* Nov. 29, 1980, 19, 33.

53. Interview with Maurice Russell, Newton, Iowa, Jan. 21, 1986.

54. Des Moines *Sunday Register,* Aug. 5, 1984.

56. Interview with James T. Orr, Blairstown, Iowa, Jan. 2, 1987; USDA *A Time to Choose,* 117-23.

Family

1. For the thirties and forties, see Erven J. Long and Kenneth H. Parsons, *How Family Labor Affects Wisconsin Farming,* Wisconsin AES Research Bulletin 167 (1950). A recent discussion of farm family research is found in Dalva E. Hedlund and Alan D. Berkowitz, *Farm Family Research in Perspective, 1965-77,* Departments of Rural Sociology and Agricultural Economics, Cornell University, Rural Sociology Bulletin 79 (1978). For the contribution of women to corn-belt agriculture in the twentieth century, see Deborah Fink, *Open Country, Iowa: Rural Women, Tradition and Change* (Albany: State University of New York Press, 1986), for an incisive analysis.

2. Emphasis on the rigidity of Central Valley farm work patterns is given in Goldschmidt, *As You Sow,* 80-92. For a familial orientation, see Yvonne Jacobson, *Passing Farms, Enduring Values: California's Santa Clara Valley* (Los Altos, Calif.: William Kaufmann, 1984), 127-152.

3. LeVeen, "Enforcing the Reclamation Act," 669.

4. Paul C. Rosenblatt et al., *The Family in Business* (San Francisco: Jossey Bass, 1985), chaps. 2 and 9.

5. Interview with Frances Von Glahn, Corcoran, Calif., July 7, 1984.

6. For European rural social structure and inheritance in historical perspective, see Jack Goody, Joan Thirsk, and E.P. Thompson, eds., *Family and Inheritance* (Cambridge: Cambridge University Press, 1976).

7. Social change in conservative rural communities that stressed familism is summarized in James A. Duncan, *Agricultural, Social and Educational Change in Rural Wisconsin, 1953-1973,* Wisconsin AES Report 3088 (1981), 26-27.

8. Mark Friedberger, "Handing Down the Home Place," *Annals of Iowa* 47 (1984). 518-36; Phebe Pjellstrom, *Swedish American Colonization in the San Joaquin Valley* (Uppsala, Sweden: Studie Ethnographie Upsaliensia, 1970).

9. Salamon, "Ethnic Communities and the Structure of Agriculture," 324-26; and Sonja Salamon, "Family Farming and Farm Community Character," in *Farm Work and Fieldwork: Anthropological Studies of North American Agriculture,* ed. M. Chibnik (Ithaca: Cornell University Press, 1987), 165-88.

10. Miner, *Culture and Agriculture,* 27. James A. Duncan, "The Relationship of Selected Cultural Characteristics of Acceptance of Educational Programs and Practices among Certain Rural Neighborhoods in Wisconsin" (Ph.D. diss., University of Wisconsin, 1953), 27-38.

11. Deed Book 51:494, 100:443, 104:7, 136:419, Fayette County Recorder's Office, West Union, Iowa.

12. Farm Account Book in the possession of Sarah Gruhn, Auburn, Iowa.

13. Ibid.

14. Interview with Sarah Gruhn, Auburn, Iowa, Nov. 10, 1982.

15. Deed Book 70:491, 72:18, 76:429, Chickasaw County Recorder's Office, New Hampton, Iowa.

16. Interview with George Gruhn, Aug. 29, 1983, Lawler, Iowa; Deed Book 129:88, 123:247, 177:158, Chicasaw Recorder's Office, New Hampton, Iowa.

17. Lorraine Kuennen, ed., *Wenger Genealogy* (Auburn, Iowa: privately printed, 1980).

18. Interview with Leo Wenger, Auburn, Iowa, Dec. 9, 1982.

19. Interviews with Glen Paulsen, Maynard, Iowa, June 10, July 7, 1985.

20. Interviews with Urban Berkes, Oelwein, Iowa, July 15, 1985, Carl Oldenburg, West Union, Iowa, May 28, 1985, and Bill Ruff, Westgate, Iowa, July 17, 1985.

21. Interview with Ron Dietrich, Belle Plaine, Iowa, July 10, 1983.

22. Deed Book 12:318-20, Benton County Recorder's Office, Vinton, Iowa.

23. Deed Book 91:505, ibid.

24. Interview with George Kettering, Jr., Van Horne, Iowa, Aug. 27, 1983.

25. Deed Book 110:57, 125:149, 609, Benton County Recorder's Office; Probate Docket 11515, Benton County Clerk of Court.

26. Interview with Erven Kettering, Van Horne, Iowa, Aug. 27, 1983.

27. Quoted in Manchester *Guardian Weekly,* July 24, 1983, 17.

28. *Using Grain Futures in the Farm Business,* Iowa AES Circular Pm-687 (1981).

29. Interview with Johnnie Gruber, Kane, Iowa, July 8, 1983.

30. Interview with Don Gruber, Kane, Iowa, Jan. 4, 1985.

31. Interview with Johnnie Gruber, Aug. 30, 1985.

32. BAE, "Rural Trends in Tulare County," 78.

33. Salamon, "Ethnic Communities and the Structure of Agriculture," 325; Rohwer, *Family Factors in Tenure Experience,* 829-30.

34. Graves, "Immigrants in Agriculture," 23-44, 71-116.

35. Interviews with Tony Soares, Hanford, Calif., May 2, 1984, Germano Avila, Hanford, Calif., Dec. 10, 1984.

36. Interview with Manuel Garcia, Hanford, Calif., Dec. 10, 1984, Oct. 20, 1985.

37. Ibid., Oct. 20, 1985.

38. Interview with Manny Cordoza, Jr., Hanford, Calif., July 20, 1984.

39. Interviews with Manny Cardoza, Sr., Hanford Calif., July 20, 1984, April 11, 1986.

40. Interviews with John DeGraad, July 5, 1984, Oct. 12, 1985, Hanford, Ca. See also Harry Cline, "Two Crops from One Field," *California-Arizona Cotton* (July 1985): 4-8, which discusses the family's double-cropping practices.

41. *Stabilizing Temporary Farm Labor Supply and Employment through Year Round Crews,* Giannini Foundation Information Series in Agricultural Economics 63-1 (1963).

42. See Hanford *Journal,* Dec. 16, 1951, for family involvement in water politics. For information on extension work with farm labor families, which Mrs. Warner directed, see Committee to Survey the Agricultural Labor Resources of the San Joaquin Valley, Hearings Held at Corcoran, Calif., Aug. 3, 1950 (Fresno Public Library Local History-Collection, mimeographed), 12-18.

43. Interview with John Warner, Hanford, Calif., Dec. 16, 1984.

44. Interview with Julie Warner, Hanford, Calif., April 12, 1986.

45. Probate Docket 167, June 13, 1903, Kings County Clerk of Court, Hanford, Calif.

46. Probate Docket 1179, Oct. 22, 1920, ibid.

47. See Hanford *Journal,* July 25, 1951, for Peter Hutton's leadership role.

48. Probate Docket 10136, Kings County Clerk of Court; Interview with Dave O'Neill, Hanford, Calif., Oct. 15, 1985.

49. Tom Slayton, interview, in "California Odyssey: The 1930's Migration to the Southern San Joaquin Valley," Oral History Archive, California State College at

Bakersfield. For the Slayton murder tragedy, see Fresno *Bee,* Sept. 16, 1984, sec. A; David Drum, "Making a Better Planter or Mulcher or Applicator," *Cotton Farming* (Jan. 1984): 13-15, is an interview with Ron Slayton published after he was murdered. James Noble Gregory, "The Dust Bowl Migration and the Emergence of an Okie Subculture in Ca., 1930-1950" (Ph.D. diss., University of California, Berkeley, 1983), 109-54, does not consider success stories in the valley.

50. Sidney T. Hardy, "A Life in Western Water Management," 114, State Water Resources Center, University of California, Regional Oral History Office, Bancroft Library, Berkeley.

51. Leake and Barnes, *Pine Flat Project on the Kings River,* provides a useful summary of the history and farming conditions of Corcoran.

52. Corcoran *Journal,* 30 September 1976, sec. E.

53. Interview with Louis Jensen, Corcoran, Calif., Sept. 21, 1984.

54. Probate Docket, 5102, Kings County Clerk of Court, gives details of the operation in 1954.

55. Interview with Donald Eaton, Stratford, Calif., Dec. 14, 1984.

56. Central coast ranchers were analyzed by Elvin Hatch, *Biography of a Small Town* (New York: Columbia University Press, 1979), 128.

57. Probate Docket 11515, Kings County Clerk of Court.

58. Interview with Tony Ingram, Lemoore, Calif., May 15, 1984.

59. Quoted in Hanford *Sentinel,* Jan. 18, 1985.

60. Interview with Tony Ingram, Oct. 22, 1985.

Community

1. Brown, *Story of Kings County,* 58.

2. Goldschmidt, *As You Sow,* 186-220.

3. *Directory of Fayette County* (Dubuque: Telegraph Herald, 1907), 182-89; *Polk's Oelwein City Directory, Including Fayette County* (Omaha: H. and R. Polk, 1941).

4. Harley E. Johansen and Glen V. Fuguitt, "Population Growth and Rural Decline: Conflicting Effects of Urban Accessibility on American Villages," *Rural Sociology* 44 (1979). 24-38.

5. See "Iowa's County Seats," Des Moines *Sunday Register,* May 18, 1986; Des Moines *Register,* May 19, 20, 1986.

6. See Michael N. Haynes and Alan L. Olmstead, "Farm Size and Community Quality: Arvin and Dinuba Revisited," *American Journal of Agricultural Economics* 66 (1984): 430-35. An exhaustive bibliography on the "Goldschmidt thesis" is provided by Robert L. Moxley, "Agriculture, Communities, and Urban Areas," in U.S. Congress, Joint Economic Committee, Subcommittee on Agriculture and Transportation, *New Dimensions in Rural Policy: Building upon Our Heritage* (Washington: Government Printing Office, 1986), 322-32.

7. See Leake and Barnes, *Pine Flat Project on the Kings River,* 16.

8. Goldschmidt, *As You Sow,* 2d ed. (Montclair, N.J.: Allanheld, Osman, 1978), 455-91, gives details of the struggle between the Bureau of Agricultural Economics and the California agricultural establishment during and just after World War II, with the publication of research assessing the Central Valley Project.

9. Interview with Evon Cody, Hanford, Calif. March 12, 1984.

10. Fresno *Bee,* Oct. 10, 1985.

11. Census manuscripts, 1910, NA.

12. *Auburn Church, 1855-1980: 125 Years* (Auburn, Iowa: n.p., 1980), 11.

13. *Kane, Iowa. The First 100 Years, 1881-1981* (Kane: Cenntenial Committee, 1981), 5.

14. Ibid., 16.

15. Census Manuscripts, 1910, NA.

16. *Auburn Church*, 18-25. See Loras C. Otting, "Gothic Splendor in Northeast Iowa," *Annals of Iowa* 48 (1986). 158, for the fund-raising incident.

17. *Auburn Church*, 26-42. As late as 1981 a reporter from the Des Moines *Register* described the community as "clannish." See Des Moines *Sunday Register*, Jan. 25, 1981.

18. *Kane, Iowa*, 39-42, 55-57.

19. Mark W. Friedberger, "Cornbelt and River City: Social Change in a Midwest Community, 1870-1930" (Ph.D. diss., University of Illinois, Chicago Circle, 1973), 244-91.

20. *Kane, Iowa*, 41.

21. Ibid., 65-69.

22. Interview with Dolores Kuennen, Auburn, Iowa, Oct. 5, 1983.

23. *Auburn Church*, 18-27.

24. Ibid. See Waucoma *Echo*, Oct. 6, 1983, for a school board election where the issue of curriculum was raised.

25. *Kane, Iowa*, 43.

26. Interview with John Blong, Auburn, Iowa, March 7, 1986.

27. Interview with Nick Kuennen, Auburn, Iowa, March 6, 1985.

28. Interview with Blong.

29. Interview with Ron Kuehn, Kane, Iowa, April 2, 1986.

30. Benton County *Union*, Feb. 11, 1985.

31. Cedar Rapids *Gazette*, Sept. 10, 1985.

32. Interview with Jim McMahon, Kane, Iowa, April 2, 1986.

33. Cedar Valley *Times*, Oct. 4, 1984.

34. Interview with Dave Bealer, Kane, Iowa, April 2, 1986.

35. Interview with William Voelker, Kane, Iowa, Aug. 10, 1983.

36. Interview with Mary Striek, Kane, Iowa, April 2, 1986.

37. Interview with James Reynolds, Hanford, Calif., March 2, 1984.

38. Interview with Harlan Hagen, Hanford, Calif., Sept. 28, 1984.

39. Lemoore Naval Air Station, *Air Installation Compatible Use Study* (Feb. 1970); Claude E. Elias, Jr., "The Impact of the Lemoore Air Base Development on Real Estate uses, Values, and Taxation" (1958), mimeographed), in Kings County Public Library, Hanford, CA.

40. Kings County Planning Commission, *Kings County General Plan* (Hanford, Calif.: KCPC, 1965), 3.

41. Interview with Cody.

42. Fresno *Bee*, July 25, 1982, sec. C.

43. One of the few studies of recent Mexican immigration and its impact on both the rural California and rural Mexican countryside is Richard Mines and Carole Frank Nuckton, *The Evolution of Mexican Migration to the United States: A Case Study*, Giannini Foundation Bulletin 1902 (1982), 16-19.

44. Interview with Doris Dooley, Hanford, Calif., Sept. 25, 1984.

45. Interview with Leonard Simmons, Hanford, Calif., July 10, 1984. See Hanford *Sentinel*, Feb. 27, 1985, for views of Hanford FFA members on farming as a career. For the success of the organization, see Fresno *Bee*, April 27, 1986.

46. Fresno *Bee*, Sept. 16, 1984, sec. A.

47. Hanford *Sentinel*, March 7, 1984.

48. Fresno *Bee*, Sept. 16, 1984, 8, 9.

49. Ibid. See Hanford *Sentinel*, June 20, 1985, for editorial opinion on the guilty verdict of one of the contract killers.

50. Interview with Harold Gibson, Hanford, Calif., July 12, 1984.

51. Interview with Lyman Griswold, Hanford, Calif., July 24, 1984.

52. Hanford *Journal*, July 26, 1951.

53. Civil Case 42660, Tulare County Clerk of Court, Visalia, Calif.; Hanford *Journal*, Dec. 14, 1951.

54. Hanford *Sentinel*, Dec. 14, 1951.

55. Hanford *Journal*, Feb. 19, 1953.

56. Interview with Gerald Schumacher, Hanford, Calif., Oct. 2, 1984.

57. Interview with William Frey, Corcoran, Calif., Sept. 30, 1984.

58. San Fransisco *Examiner*, July 28, 1947, 25.

59. Corcoran *Journal*, Sept. 30, 1976, sec. C.

60. Ibid., sec. B.

61. Ibid., sec. E.

62. One of the few articles on the Payne family appeared in the Los Angeles *Times*, Nov. 18, 1979.

63. Official Records, Kings County Recorder's Office, Hanford, Calif., Book 54:409, 419, 421, Book 69:297, 479. Between 1968 and 1970, Paynes received $12 million in crop subsidies; in 1970 the company was the fifth largest beneficiary of the California Land Conservation Act, which reduced Payne's property assessments by $3.5 million. See Merrill, "American Involvement and the Resurgence in the Australian Cotton-Growing Industry," 149.

64. Senate Committee on Education and Labor, *Hearings to Investigate Violations of Free Speech and Assembly and Interference with the Right to Organize and Bargain Collectively*, 26423.

65. Committee to Survey the Agricultural Labor Resources of the San Joaquin Valley, Hearings, 8-11.

66. Ibid., 21.

67. Ibid., 17.

68. Ibid., 80.

69. Kings County Planning Commission, "Staff Report on Review of Corcoran Area General Plan" (1965), 2.

70. Interview with Cody.

71. See Los Angeles *Times*, March 27, 1983.

72. Civil Case 21140, memo marked "Confidential from American Investigation Service," Kings County Clerk of Court.

73. Los Angeles *Times*, Nov. 17, 1979.

74. Interview with Forrest Mitchell, Visalia, Calif., Oct. 13, 1985.

75. Hanford *Sentinel*, April 11, 1985; City of Corcoran Redevelopment Agency, *Corcoran, California: Our Spirit Will Move You* (n.d.).

76. Hanford *Sentinel*, Nov. 20, 1984; Corcoran *Journal*, Sept. 19, 1985. The Corcoran *Journal*, Oct. 17, 1985, claimed the prison would employ inmates in a textbook plant, an industrial laundry, dairy, and furniture shop.

77. See Engineering Economics Association, *Economic and Fiscal Impacts of the Proposed State Prisons in Kings County, Ca.* (Berkeley. EEA, 1985); *Cal. State Prison Corcoran: Environmental Assessment Study* (Sacramento: State of California, 1985).

Corn-Belt Crisis

1. The unfortunate John Block received a good deal of this hostility. Once, the Des Moines *Sunday Register* published a full list of Block's fifty-four farm loans, which totaled $6.7-$10 million. The disclosure took up a full page in the June 10, 1984, edition. See also the Des Moines *Register*, Sept. 7, 1985, for the grilling of the farm credit administrator Donald Wilkinson by Senator Charles Grassley.

2. The details of the Extension Service's mobilization, and the major components of the Assist program, can be found in Iowa State University, College of Agriculture,

Department of Economics, "Progress Report: Some Perspectives on Farm Financial Stress" (Sept. 1984), 63-64. It is important to point out that the criticism leveled at the Extension Service in the early stages of the farm crisis, was a reflection of perceptions gathered in traveling around the state early in 1985. As early as 1982, some extension personnel saw the warning signs of a downturn; however, little public action was taken until two years later. See, for example, the Des Moines *Sunday Register,* March 25, 1984, for details on the Extension Service's extra effort with management seminars in the fall of 1983. See also Neil Harl's many warnings about the state of the agricultural economy and the need for debt restructuring in, for example, the Des Moines *Sunday Register,* June 10, 1984.

3. Thomas A. Lyson, "Pathways into Production Agriculture: The Structuring of Farm Recruitment in the United States," in *Research in Rural Sociology and Development,* ed. Harry K. Schwarzweller, (Greenwich, Conn.: Jai Press, 1984), 1:79-104.

4. Des Moines *Register,* Dec. 19, 1985; Robert W. Jolly et al., "Incidence, Intensity, and Duration of Financial Stress among Farm Firms" (Paper presented at the 1985 American Agricultural Economics Association, Ames, Iowa, Aug. 4, 1985) 5.

5. ISU, "Progress Report," 43.

6. *Economist* (London), Dec. 1, 1984, 34.

7. Raup, "What prospective Changes May Mean," 39.

8. Ibid., 38.

9. See Des Moines *Register,* June 16-21, 1985, for the impact of the farm crisis on the northwest Iowa county of Pocahontas.

10. Roger G. Ginder et al., "Impact of the Farm Financial Crisis on Agribusiness Firms and Rural Communities," *American Journal of Agricultural Economics* 36 (1985): 1187.

11. See *California Farmer,* Feb. 15, March 22, 1986, for the impact of the farm crisis in Newman and Modesto, Calif.

12. Ginder, "Impact of Farm Financial Crisis," 1189.

13. Iowa Cooperative Extension, *Analysis of Farmers Leaving Agriculture for Financial Reasons,* Pm-1207 (1985), 1-6.

14. Interviews with John Higgins, Galva, Iowa, Dean Ruser, Holstein, Iowa, Dec. 20, 1986.

15. See Des Moines *Sunday Register,* April 15, 1984, for the details of the founding. Interview with the Reverend David Ostendorf, Des Moines, Iowa, Jan. 3, 1985. The American Agriculture Movement was founded in 1977 to promote parity in farm commodity production. Between 1977 and 1983, the AAM, the United States Farmers' Association, the Farmers' Union, and other organizations conducted various actions around the country to protest government agricultural policy and decisions by lenders such as the PCA and Farmer's Home Administration. Most of this activity occured in Colorado, Ohio, Illinois, North Dakota, and Minnesota. See Fite, *American Farmers,* 209-17. For a discussion of this period by a visiting Canadian farm activist, see Allen Wilford, *Farm Gate Defense* (Toronto: N.C. Press, 1984), chaps. 13-15.

16. Jenkins, *Politics of Insurgency,* 212.

17. Interview with Fr. Norman White, Dubuque, Iowa, Jan. 25, 1985.

18. Stephen Wright, "Digging In," *Esquire* (Dec. 1984). 514.

19. Ibid., 510.

20. Des Moines *Register,* Sept. 27, 1984.

21. Des Moines *Register,* July 2, 1986.

22. Des Moines *Sunday Register,* April 15, 1984.

23. See Des Moines *Sunday Register,* Jan. 20 1985, editorial "Hysteria over Farm Crisis"; and Des Moines Register, Jan. 21, 1985, for the uncertain political climate created by the farm crisis; interview with Ostendorf, March 13, 1986.

24. Personal observations in Des Moines, Jan. 21, 22, 24, 1985, as well as meetings in

Atlantic, Jan. 23, 1985, Holstein, Feb. 21, 1985, and Belmond (a Catholic Rural Life sponsored meeting), Feb. 26, 1985.

25. For clergy mobilization, see Des Moines *Sunday Register,* Oct. 27, 1985; interview with Ostendorf, March 13, 1986.

26. Prairiefire, "Farm Credit: Warning Signs of Liquidation" (1985, mimeographed).

27. Charles Isenhart, "The Fields of Battle," *Catholic Rural Life* (Nov. 1985): 5, captures the emotionally charged atmosphere brought on by the farm crisis in northeast Iowa in the spring of 1985.

28. Ibid., 5-6.

29. Interviews with Deb and Don Bahe, Stanley, Iowa, June 5, 1985, Jim Koch, West Union, Iowa, March 10, 1985, Loren Osmundson, West Union, Iowa, July 7, 1985.

30. Quoted in Des Moines *Register,* Jan. 22, 1986.

31. Prairiefire, "Farm Credit," summarizes the arcane regulations succinctly.

32. Des Moines *Sunday Register,* May 28, 1984.

33. Richard Stageman's statement before Congress was reprinted in American Bar Association, "How Lawyers Can Help Lenders and Borrowers" (Seminar in Agricultural Finance, Denver, March 20-21, 1985); Des Moines *Register,* Nov. 7, 1985; Bankruptcy Court, Northern District of Iowa, Cedar Rapids.

34. Des Moines *Register,* May 26, 1985.

35. Ibid.

36. Prairiefire Rural Action, *Who Is behind the Farm Crisis* (Des Moines, Iowa, 1985).

37. Des Moines *Register,* Sept. 29, 1985. See also Michael Scholer, "A Right Wing Overview, "Catholic Rural Life (Nov. 1985), 9-10.

38. Des Moines *Register,* July 2, 1985.

39. Ibid., June 2, 1985.

40. *Agri Finance* (Jan. 1986), 60.

41. I watched this incident in Fayette, Iowa, June 26, 1985.

42. Interview with Mike Thompson, Iowa Farmer/Creditor Mediation Service, Des Moines, Iowa, March 13, 1986. The Iowa legislature funded mediation in the spring of 1986, thus officially sanctioning this method of resolving lender-farmer deadlock.

43. Mark Singer, *Funny Money* (New York: Alfred A. Knopf, 1985), chronicles the rise and fall of Penn Square Bank and the boom and bust in the oil business in Oklahoma.

44. Michael Boehlje, "Financial Stress in Agriculture: Implications for Farmers, Lenders and Consumers," testimony presented before the House Subcommittee on Economic Stabilization of the Committee on Banking, Finance and Urban Affairs, *Problems of Farm Credit,* 99th Cong., 1st Sess. (Washington D.C.: Government Printing Office, 1986), 82-84.

45. I attended these meetings, in Holstein, Feb. 21, 1985, and Fontanelle, Jan. 22, 1985.

46. Quoted in *California Farmer,* May 3, 1986, 11.

47. Senate Subcommittee on Administrative Practice and Procedure of the Committee on the Judiciary, *Farm Credit Crisis,* 99th Cong., 1st Sess. (Washington D.C.: Government Printing Office, 1986), 72, and see 73-94 for materials from Hawkeye Bank Corporation, which document a lack of regulator forbearance on farm loans in the portfolios of member banks.

48. *Agri Finance* (March 1985), 24-25. For the Prairiefire "Bank Response Team," see Jerry Perkins, "Target: Failing Banks," *Agri Finance* (April 1986), 19.

49. Senate Subcommittee on Administrative Practice and Procedure, *Farm Credit Crisis,* 145-46.

50. Ibid., 147.

51. Ibid., 210. The hearing was considered important enough to be broadcast live over radio station WOI, until the station went off the air at sundown. No current Iowa

employees of either the Land Bank or the PCA were asked to testify, prompting at least one director to complain to Senator Grassley. "The system," he wrote, "has many hardworking and dedicated employees who are working diligently and long hours to try to keep farm families on the farm, while at the same time make some business and lending decisions which are in the best interest of first, the farmer, then the Association, then the investing public." As to the association's lack of forbearance, this director invited the senator to a loan committee decision session, "so that you may gain first hand knowledge on the thought process involved, research conducted, and the humanitarianism used in our decision making process before foreclosures are initiated" (ibid., 304).

52. Ibid., 172.

53. Interview with Maurice Russell, Newton, Iowa, May 2, 1985.

54. Senate Subcommitte on Administrative Practice and Procedure, *Farm Credit Crisis*, 245.

55. Des Moines *Sunday Register*, Jan. 12, 1986.

56. Des Moines *Register*, Jan. 3, 1986.

57. On the Hills murder, see Des Moines *Register*, Dec. 10, 11, 1985, and *Sunday Register*, Dec. 15, 1985.

58. Interview with Russell.

59. Obviously, state government—the Department of Agriculture, the governor, the legislature—were very much part of the establishment, but because of their physical distance from and lack of contact with the average farm family, I made no effort to analyze their role.

60. Interview with Fran Philips, Rural Concern, Des Moines, Iowa, June 6, 1986.

61. Iowa State University Cooperative Extension, Rural Concern Statistical Information, March, May 1986.

62. See Des Moines *Sunday Register*, Dec. 8, 1985, for an interview with Iowa Farm Bureau head Dean Kleckner. I interviewed Kleckner in Charles City, Iowa, Feb. 23, 1985.

63. I attended some of these workshops, in Vinton, Feb. 11, 1986, Cedar Rapids, Jan. 26, 1986, and Charles City, Feb. 23, 1985.

64. Ironically the University of Iowa closed its Agricultural Law Center in the 1970s because, in the words of one attorney, its Ivy League–trained faculty "looked down their noses at Hog law." Interview with Neil Hamilton, Drake University, Cedar Falls, Iowa, July 22, 1986. See *Iowa Farmer Today*, Feb. 9, 1985. One lawyer called Chapter 11 "probably the finest piece of legislation I have ever worked with." However, as early as May 1984, some bankers were questioning the utility of bankruptcy as a strategy. See Des Moines *Sunday Register*, May 28, 1984.

65. Interview with Tim Jackson, staff lawyer, Iowa State University ASSIST program, Waterloo, Iowa, Feb. 22, 1986. For the activities of one attorney who, with over two hundred farmer cases pending, surrendered his license and left the state, see Des Moines *Sunday Register*, July 20, 1986. See also cases 8500779W, and 8500839, U.S. Bankruptcy Court, Northern District of Iowa, Cedar Rapids, for client petitions for the reduction of attorney fees.

66. This material was taken from a speech given by attorney Fred Dumbaugh, Brooklyn, Iowa, March 14, 1986.

67. Ibid.

68. Ibid.

69. Ibid.

70. See Des Moines *Register*, Jan. 4, June 2, 1985, March 28, 1986.

71. Interview with Ostendorf, Jan. 3, 1985.

72. Interviews with Peter Brent, Des Moines, Iowa, Oct. 29, 1985, Ostendorf, March 13, 1986. See Des Moines *Register*, April 16, 1986. Throughout 1983-86 Governor Terry Branstad found himself in a difficult position politically and personally. A conservative

Republican (he was once active in Young Americans for Freedom), he identified with the Reagan administration, whose positions on agriculture were unpopular in Iowa by December 1984. In addition, the governor was a farmer, whose financial position was jeopardized by the farm crisis. Tax records showed that his farming operation lost over thirty thousand dollars in 1985, while his debt to asset ratio was 93 percent. This placed him in the category of farmers under "considerable financial stress." In addition, his brother and his cousin, an Iowa legislator, were also in financial trouble on the farm.

73. See Des Moines *Register,* Dec. 24, 1985, for the frustrating efforts of one county mental health director in Fayette County, Iowa, to make contact with farmers through an open house at his clinic. Interview with Pete Zevenbergen, Cedar Rapids, Iowa, April 26, 198´

74. See Iowa State University Cooperative Extension Service, *Self Help Groups: Neighbor to Neightor,* FE-F-270a (1985), 1-8. For the effectiveness of one county Extension agent in Sac County, Iowa, see Washington *Post,* Dec. 15, 1985.

75. Personal communication with Dolores Fagel, Randalia, Iowa, March 31, 1986.

76. Interview with Julie Paulsen, Cedar Rapids, Iowa, April 26, 1986. See also Des Moines *Register,* Feb. 17, 1986, for a farmer-to-farmer program with high-school-age children.

77. Interview with Patrick Singel, director, Northwest Iowa Mental Health Center, Spencer, June 3, 1986.

78. Joan Blundall, "Community and Family: Responding to Immediate Needs" (Paper presented to Iowa Governor's Conference on Agriculture in Transition, May 1985), 7.

79. Joan Blundall, "Support Groups and the Rural Crisis" (Unpublished paper, Northwest Iowa Mental Health Center, Spencer, 1985), 1.

80. Joan Blundall, "Peer Listening Program: Empowering Community Members as Helpers" (Unpublished paper, Northwest Iowa Mental Health Center, Spencer, 1985), 2-3.

81. Des Moines *Sunday Register,* Oct. 27, 1985.

82. Ibid.

Central Valley Crisis

1. Hanford *Sentinel,* March 1, 1985. See also San Fransisco *Chronicle,* Oct. 8, 1984, for a confident prediction that California would escape the downturn.

2. Fresno *Bee,* March 16, 1985.

3. Ibid., May 12, 1985.

4. See articles from the UPI Reporter Sonja Hilgren on the midwestern farm crisis, reprinted in the Hanford *Sentinel,* Feb. 27, March 1, 6, 1985.

5. Interviews with Stuart Bartlett, Corcoran, Calif., Oct. 16, 1985, Joe Cotta, Laton, Calif., Oct. 22, 1985.

6. For the failure of large enterprises, See Sacramento *Bee,* Sept. 29, 1985. For evidence that large enterprises had advantages over smaller, see Fresno *Bee,* May 18, 1986. For the Payne cutbacks, see Hanford *Sentinel,* Nov. 21, 1985. For Payne's $13.7 million involvement in the cotton program, see Fresno *Bee,* 15 April 1987; Hanford *Sentinel,* 16 April 1987.

7. Interview with Cornelius Gallagher, Fresno, Calif., April 8, 1986.

8. For the diversification of the San Joaquin Valley economy, see Sacramento *Bee,* March 31, 1986.

9. Journalists found great difficulties in obtaining information about the extent of the farm crisis from farmers. Telephone interview with J.L. Levinson, Lemoore, Calif. Feb. 29, 1986; Interview with Jeanie Borba, Fresno, Calif., April 7, 1986.

10. Interview with Gallagher.
11. Quoted in the Lemoore *Advance,* June 20, 1985.
12. Fresno *Bee,* May 12, 1985.
13. Sacramento *Bee,* Sept. 29, 1985.
14. Lemoore *Advance,* June 27, 1985.
15. Ibid., June 20, 1985.
16. *California Ag Alert,* Oct. 2, 1985, 19.
17. Interview with Eleanor Coehlo, Hanford, Calif., April 9, 1986.
18. Fresno *Bee,* May 12, 1985.
19. Interview with Julie Warner, Hanford, Calif., April 12, 1986.
20. Telephone interview with Tony Oliveira, Lemoore, Calif. Feb. 17, 1986.
21. Interview with Riley Forrest, Corcoran, Calif., April 10, 1986.
22. Fresno *Bee,* April 12, 1986.
23. San Fransisco *Examiner,* March 17, 1986.
24. Ibid.
25. Fresno *Bee,* May 12, 1985; San Fransisco *Examiner,* March 20, 1986.
26. San Fransisco *Examiner,* March 18, 1986; telephone interview with Forrest Mitchell, Corcoran, Calif., Dec. 29, 1986.
27. See *California Farmer,* Jan. 18, 1986, for an explanation of commercial lenders' credit strategy during the boom.
28. *National Journal,* March 29, 1986, 771.
29. Interview with Gallagher, Fresno, Oct. 15, 1985; Lemoore *Advance,* June 27, 1985.
30. San Fransisco *Examiner,* March 18, 1986.
31. Ibid.
32. Bank of America, *Agricultural Loan Preparation Guide* (Sept. 1985).
33. Quoted in Sacramento *Bee,* March 18, 1986.
34. San Fransisco *Examiner* March 9, 16, 1986; *California Farmer,* April 7, 1986.
35. Interview with Gallagher, Fresno, April 7, 1986.
36. Interview with Don Edwards, Hanford, Calif., April 9, 1986.
37. Reel 1359, 267-350, Official Records, Kings County Recorder's Office, Hanford, Calif., *California Farmer,* Oct. 18, 1986; Fresno *Bee,* July 19, 1987.
38. USDA, Extension Service, *Cooperative Extension and Agricultural Profitability—Intensive Assistance for Financially Distressed Farmers* (Nov. 1985); "How Are Extension and Research Responding to the Farm Crisis?" *Rural Development News* 10 (1986): 1-7.
39. See, for example, William H. Friedland and Tom Keppel, *Production or Perish: Changing the Inequalities of Agricultural Research Priorities* (Santa Cruz, Calif.: University of California, 1979), 5-10. For a critical assessment of the role of the College of Agriculture at the University of California, see Emmett Preston Fiske, "The College and Its Constituency: Rural and Community Development at the University of California, 1875-1978" (Ph.D. diss., University of California, Davis, 1979).
40. Des Moines *Sunday Register,* Sept. 15, 1985.
41. The list was published in *California Agriculture* (Feb. 1980): 18-20.
42. Interview with Robert Beede, Hanford, Calif., April 9, 1986; Nancy Dickinson et al., "Assessing the Needs of Families in Transition" (Unpublished paper kindly supplied by the author).
43. *California Farmer,* Feb. 15, 1986; "In California," *Catholic Rural Life* (Feb. 1986): 5-8.
44. Interview with Cindy Bauer, Fresno, Calif., April 8, 1986; telephone interview with Nancy Dickinson, University of California Extension, Berkeley, Dec. 29, 1986.
45. Interview with Jan Kahn, April 11, 1986.
46. Fresno *Bee,* May 18, 1986.
47. Ibid., March 30, 1986; Sacramento *Bee,* March 24, 1986.
48. *California Farmer,* May 2, 1987.

49. For a technical discussion of the problem of selenium, see Richard G. Bureau, "Environmental Chemistry of Selenium," *California Agriculture* (July-Aug. 1985): 16-18. For a historical treatment, see Robert L. Kelley and Ronald L. Nye, "Historical Perspectives on Salinity and Drainage Problems in Ca.," *California Agriculture* (Oct. 1984): 4-6.

50. Fresno *Bee*, March 16, 1985.

51. Gale Norman, "Reckoning Day Has Arrived," *California-Arizona Cotton* (Jan. 1985): 10-18; Fresno *Bee*, March 16, Aug. 11, 1985.

52. *California Ag. Alert*, April 2, 1986.

53. Gale Norman, "West Side Drainage Alliance," *California-Arizona Cotton* (Feb. 1986): 4-5. For a pessimistic appraisal of the selenium issue in the long term, see Sam Wilson, "Westlands Beats the Deadline: What Now?" *California Farmer*, June 21, 1986.

54. Telephone interview with Walter Shubin, Kerman, Calif., Sept. 7, 1986.

CONCLUSION

1. Congress, Office of Technology Assessment, *Technology, Public Policy, and the Changing Sstructure of Agriculture* (Washington, D.C.: Goverment Printing Office, 1986), 285. Jack Doyle discusses agribusiness involvement in high-tech agriculture and food supply in *Altered Harvest: Agriculture, Genetics, and the Fate of the World's Food Supply* (New York: Viking, 1985).

2. Congress, *Technology, Public Policy, and the Changing Structure of Agriculture*, 286.

3. William L. Flinn and Frederick H. Buttel, *Sociological Aspects of Farm Size: Ideological and Social Consequences of Scale in Agriculture*, Cornell Rural Sociology Bulletin 114 (1980), 28-29.

4. A joke making the rounds of small-town cafes in Iowa sums up the situation: "Have you heard about the latest case of child abuse? The father gave the son the farm." Quoted in Des Moines *Sunday Register*, March 9, 1986.

5. Center for Rural Affairs, *Small Farm Resources Project* (Hartington, Nebr., n.d.); Small Farm Resources Project, *What Is Appropriate Alternative Agriculture?* (Hartington, Nebr.: Center for Rural Affairs, n.d.).

6. California Association of Family Farmers, *Farm Link* (May/June 1986), 2.

Index